Higher Tier

2nd Edition

OCR GCSE Mathematics

Revision Guide

Series Editor: Brian Seager
Authors: Howard Baxter, Michael Handbury, John Jeskins, Jean Matthews, Mark Patmore

HODDER EDUCATION
PART OF HACHETTE LIVRE UK

Although every effort has been made to ensure that website addresses are correct at time of going to press, Hodder Education cannot be held responsible for the content of any website mentioned in this book. It is sometimes possible to find a relocated web page by typing in the address of the home page for a website in the URL window of your browser.

Hachette Livre UK's policy is to use papers that are natural, renewable and recyclable products and made from wood grown in sustainable forests. The logging and manufacturing processes are expected to conform to the environmental regulations of the country of origin.

Orders: please contact Bookpoint Ltd, 130 Milton Park, Abingdon, Oxon OX14 4SB. Telephone: (44) 01235 827720. Fax: (44) 01235 400454. Lines are open 9.00–5.00, Monday to Saturday, with a 24-hour message answering service. Visit our website at www.hoddereducation.co.uk

© Howard Baxter, Michael Handbury, John Jeskins, Jean Matthews, Mark Patmore, Brian Seager, Eddie Wilde, 2002, 2003, 2004, 2008

First published in 2002, 2003, 2004 by
Hodder Education,
Part of Hachette Livre UK
338 Euston Road
London NW1 3BH

This second edition first published 2008

Impression number   5 4 3 2 1
Year   2013 2012 2011 2010 2009 2008

Cover photo © Mark Harwood/Getty Images
Typeset in 10pt Futura Book by Tech-Set Ltd, Gateshead
Printed in Spain

A catalogue record for this title is available from the British Library.

ISBN-13: 978 0 340 95917 6

Introduction

This book is intended to help you prepare for the final examinations of the Higher Tier of GCSE Mathematics. The book covers all you need to know for the terminal examination.

The topics from the specification have been left in their original groups: Number; Algebra; Shape and space; Handling data. Each topic has been introduced in the same way: a reminder of definitions and techniques: a Test Yourself question, with the solution at the bottom of the page; frequently, a Chief Examiner Says tip to give further guidance and help you tackle examination questions. More practice is available, after each Test Yourself question, on the CD. At the end of each topic there is an exam question and its solution. This is followed by further exam questions for you to attempt. There is also further exam practice on the CD. The answers to all the extra questions can be found on the CD.

How to use this book

There are many ways of using the book. You could use it for reference when using a different book. You could use it by opening a page randomly and checking that you can do the questions. The most satisfactory way, however, will be to use it systematically as part of your planned revision. Whether you choose to work through it in order or jump between topic areas, you can keep a check on your progress by using the Revision records at the end of the book.

Once you have chosen a topic, read the reminders and try the Test Yourself question. If you get it right, you can go on to the next. If not, look at where you went wrong and then work through the More Practice questions. If you got it right but are not very confident, do some or all the More Practice questions. At the end of the topic, try the exam question without looking at the solution. When you have understood that, do all the More Exam Practice questions – this is important as this will reinforce your learning. Finally, don't forget to tick the box on your Revision record.

You will notice that this book is divided into two sections. Section 1 deals with topics you can expect to meet in the early part of an examination paper – those questions which earn the lower grades. Section 2 covers the topics from the later parts of the paper, which must be answered to achieve the higher grades. You may find that you can work through Section 1 fairly quickly. If so, do not be tempted to skip it, as you need to get much of the early work right if you want to obtain a higher grade as it is the total marks that count and a high grade will need high marks.

If you know that your knowledge is worse in certain topic areas, don't leave these to the end of your revision programme. Put them in at the start so that you have time to return to them nearer the end of the revision period.

 This symbol next to a question means you are not allowed to use a calculator. These questions would appear in a non-calculator paper or section of a paper.

 This symbol next to a question means you need to use a calculator. These questions would appear in a calculator paper or section of a paper.

If a question has neither symbol, you can choose whether or not to use your calculator but it is unlikely to be required to answer the question.

Examination tips

There will be a formulae sheet on the second page of the exam papers. It contains several formulae. Make sure you know what they are. Everything else you will have to remember.

Make sure you read the instructions carefully, both those on the front of the paper and those in each question. Here are the meanings of some of the words used.

- **Write, write down, state** – little working out will be needed and no explanation is required.
- **Calculate, find** – something to be worked out, with a calculator if appropriate. It is a good idea to show the steps of your working as this may earn marks even if the answer is wrong.
- **Solve** – show all the steps in solving the equation.
- **Prove, show** – all the steps needed, including reasons, must be shown in a logical way.
- **Deduce, hence** – use a previous result to help you find the answer.
- **Draw** – draw as accurately and carefully as you can.
- **Sketch** – need not be accurate but should show the essential features.
- **Explain** – give(a) brief reason(s). The number of features you need to mention can be judged by looking at the marks. One mark probably means only one reason is required.

As well as this book, there are a lot of websites that will help you revise. Go to http://www.m-a.org.uk/links/revision for a list of links.

Brian Seager, 2008
Series editor

Contents

Integers

Prime factors

- You need to be able to write numbers as the product of their prime factors. For example, $12 = 2 \times 2 \times 3 = 2^2 \times 3$.
- There are different methods of finding the product of the prime factors. Use whichever method you know.

Chief Examiner Says

Remember that prime numbers are numbers that have only two factors, 1 and the number itself.

Test Yourself (1)

Write 48 as the product of its prime factors.

More Practice N1

Highest common factor (HCF)

- The HCF of a set of numbers is the largest number that will divide into each of the numbers.
- To find the HCF, write each number as the product of its prime factors and choose the factors that are common to all the given numbers.

Test Yourself (2)

Find the HCF of 120, 36 and 84.

Lowest common multiple (LCM)

- The LCM of a set of numbers is the smallest number that can be divided exactly by all the numbers.
- To find the LCM, write each number as the product of its prime factors and choose the largest number of each prime that appears in the lists.

Test Yourself (3)

Find the LCM of 120, 36 and 84.

More Practice N2

Solutions

Test Yourself (1)

Method 1

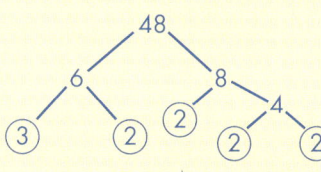

Method 2

```
2 | 48
2 | 24
2 | 12
2 | 6
    3
```

Break the numbers into factors until you end with prime numbers.

$48 = 2 \times 2 \times 2 \times 2 \times 3$ or $2^4 \times 3$

Divide by the smallest prime number, 2, as many times as you can. Then divide by the next smallest prime number, 3; then 5; then 7, and so on.

Test Yourself (2)

$120 = ②\times②\times 2 \times③\times 5$
$36 = ②\times②\times③\times 3$
$84 = ②\times②\times③\times 7$
HCF is $2 \times 2 \times 3 = 12$

2, 2 and 3 are common to all three numbers.

Chief Examiner Says

Check that 12 divides into 120, 36 and 84.

Test Yourself (3)

$120 = ②\times②\times②\times③\times⑤$

120 has the largest number of 2s (three) and 5s (one).

$36 = 2 \times 2 \times③\times③$

36 has the largest number of 3s (two).

$84 = 2 \times 2 \times 3 \times⑦$

84 has the largest number of 7s (one).

LCM is
$2 \times 2 \times 2 \times 3 \times 3 \times 5 \times 7 = 2520$

Chief Examiner Says

Check that 120, 36 and 84 all divide into 2520.

Indices

- When multiplying, add the indices.
  $n^a \times n^b = n^{a+b}$
- When dividing, subtract the indices.
  $n^a \div n^b = n^{a-b}$

Chief Examiner Says

Remember that $3^1 = 3$.

Test Yourself (1)

Write these as a single power of 3.

a) $3^2 \times 3^4$

b) $3^5 \div 3^2$

c) $3^2 \times 3^5 \div 3^3$

 More Practice N3

Here is an exam question ...

a) Find the HCF and LCM of 12 and 16. [4]

b) Work out these, writing each answer as a whole number.
  i) $5^6 \div 5^4$ [1]
  ii) $2^3 \times 2^5 \div 2^7$ [1]
  iii) $6^2 \times 5^2 \div 2^2$ [2]

Chief Examiner Says

These are different numbers so do not try to collect the indices.

... and its solution

a) $12 = 2 \times 2 \times 3$
$16 = 2 \times 2 \times 2 \times 2$
$HCF = 2 \times 2$ *Two 2s are common to both.*
$\quad = 4$
$LCM = 2 \times 2 \times 2 \times 2 \times 3$
$\quad = 48$ *Four 2s and one 3 are in at least one of the numbers.*

b) i) $5^6 \div 5^4 = 5^2$ *6 − 4 = 2*
$\quad\quad\quad = 25$

ii) $2^3 \times 2^5 \div 2^7 = 2^1$ *3 + 5 − 7 = 1*
$\quad\quad\quad = 2$

iii) $6^2 \times 5^2 \div 2^2 = 36 \times 25 \div 4$
$\quad\quad\quad = 225$

Now Try These Exam Questions

1 Write the following as whole numbers.
  a) 2^6 [1]
  b) 5^3 [1]
  c) $4^5 \times 4^2 \div 4^3$ [2]

2 a) Write 30 as the product of its primes. [2]
  b) Write down the prime factor of 30 that is also a prime factor of 21. [1]

3 Find the HCF and LCM of 10, 12 and 20. [5]

 More Exam Practice NE1

Solutions

Test Yourself (1)

a) 3^6 *2 + 4 = 6*

b) 3^3 *5 − 2 = 3*

c) 3^4 *2 + 5 − 3 = 4*

Fractions

Mixed numbers and improper fractions

- An improper fraction is a fraction with the numerator larger than the denominator.
- To change an improper fraction to a mixed number, divide the denominator into the numerator and write the remainder over the denominator as a fraction.
- To change a mixed number to an improper fraction, multiply the whole number by the denominator and add on the numerator. Then write this number over the denominator.

Chief Examiner Says

When changing from an improper fraction to a mixed number, the most common error is to put the remainder over the numerator rather than the denominator.

Test Yourself (1)

a) Change $\frac{14}{5}$ to a mixed number.

b) Change $3\frac{5}{8}$ to an improper fraction.

More Practice N4

Adding and subtracting fractions

- To add or subtract fractions, change the fractions to equivalent fractions with the same denominator, and add or subtract the numerators.
- When mixed numbers are involved, deal with the whole numbers first.
- If, in subtraction, the first fraction part is smaller than the second, change a whole number to a fraction.

Test Yourself (2)

Work out these.

a) $\frac{1}{6} + \frac{3}{4}$ **b)** $\frac{4}{5} - \frac{2}{3}$

c) $2\frac{2}{3} + 1\frac{3}{8}$ **d)** $3\frac{1}{4} - 1\frac{2}{5}$

More Practice N5

Solutions

Test Yourself (1)

a) $2\frac{4}{5}$ $14 \div 5 = 2 \text{ r } 4$

b) $\frac{29}{8}$ $3 \times 8 + 5 = 29$

Test Yourself (2)

a) $\frac{1}{6} + \frac{3}{4} = \frac{2}{12} + \frac{9}{12}$ Change each fraction to an equivalent fraction with a denominator of 12.

$= \frac{11}{12}$

Then add the numerators.

b) $\frac{4}{5} - \frac{2}{3} = \frac{12}{15} - \frac{10}{15}$ Change each fraction to an equivalent fraction with a denominator of 15.

$= \frac{2}{15}$

Then subtract the numerators.

c) $2\frac{2}{3} + 1\frac{3}{8} = 3 + \frac{2}{3} + \frac{3}{8}$ First add the whole numbers.

$= 3 + \frac{16}{24} + \frac{9}{24}$ Change each fraction to an equivalent fraction with a denominator of 24.

$= 3 + \frac{25}{24}$

$= 4\frac{1}{24}$ Add the numerators.

Change the improper fraction to a mixed number and add the whole numbers.

d) $3\frac{1}{4} - 1\frac{2}{5} = 2 + \frac{1}{4} - \frac{2}{5}$ Subtract the whole numbers.

$= 2 + \frac{5}{20} - \frac{8}{20}$ Change each fraction to an equivalent fraction with a denominator of 20.

$= 1 + \frac{20}{20} + \frac{5}{20} - \frac{8}{20}$

$= 1\frac{17}{20}$

Collect the fractions.

Change one of the whole numbers to a fraction.

Multiplying fractions

- When multiplying a fraction by a whole number, multiply the numerator by the whole number and then simplify.
- When multiplying fractions, multiply the numerators, multiply the denominators and then simplify.

Chief Examiner Says

Remember, $\frac{1}{2} \times \frac{1}{3} = \frac{1}{6}$. A common error is to multiply 1×1 and get 2.

Test Yourself (1)

Work out these, giving the answers as simply as possible.

a) $\frac{5}{6} \times 9$ b) $\frac{4}{5} \times \frac{5}{9}$

More Practice N6

Dividing fractions

- When dividing fractions, turn the second fraction upside down and multiply.

Test Yourself (2)

Work out these, giving the answers as simply as possible.

a) $\frac{3}{4} \div 6$ b) $\frac{3}{5} \div \frac{7}{10}$

More Practice N7

Here is an exam question ...   ... and its solution

Jane spent $\frac{1}{3}$ of her pay,
gave her mother $\frac{2}{5}$ of her pay
and saved the rest.

What fraction of her pay did Jane save?   **[3]**

$\frac{1}{3} + \frac{2}{5} = \frac{5}{15} + \frac{6}{15}$
$= \frac{11}{15}$

She saved $1 - \frac{11}{15} = \frac{4}{15}$

Now Try These Exam Questions

1 Work out the following, giving your answers as simply as possible.

a) $\frac{2}{3} + \frac{4}{5}$ **[2]**

b) $\frac{3}{5} \times \frac{5}{6}$ **[2]**

2 a) Put these fractions in order of size, smallest first.

$\frac{3}{4}, \frac{7}{10}, \frac{3}{5}, \frac{5}{8}$ **[2]**

b) Work out the sum of these fractions.   **[2]**

Solutions

Test Yourself (1)

a) $\frac{5}{6} \times 9 = \frac{45}{6}$ OR $\frac{5}{6} \times 9 = \frac{5}{2} \times 3$
       $= \frac{15}{2}$ $= \frac{15}{2}$
       $= 7\frac{1}{2}$ $= 7\frac{1}{2}$

> You can simplify either before or after multiplying.

b) $\frac{4}{5} \times \frac{5}{9} = \frac{20}{45}$ OR $\frac{4}{5} \times \frac{5}{9} = \frac{4}{1} \times \frac{1}{9}$
       $= \frac{4}{9}$ $= \frac{4}{9}$

Test Yourself (2)

a) $\frac{3}{4} \div 6 = \frac{3}{4} \div \frac{6}{1}$ ◀ $6 = \frac{6}{1}$
       $= \frac{3}{4} \times \frac{1}{6}$ ◀ Turn $\frac{6}{1}$ upside down.
       $= \frac{1}{4} \times \frac{1}{2}$ ◀ Cancel
       $= \frac{1}{8}$

b) $\frac{3}{5} \div \frac{7}{10} = \frac{3}{5} \times \frac{10}{7}$ ◀ Turn $\frac{7}{10}$ upside down.
          $= \frac{3}{1} \times \frac{2}{7}$ ◀ Cancel
          $= \frac{6}{7}$

3 Work out these, giving your answers as simply as possible.

   a) $2\frac{3}{8} - 1\frac{1}{2}$ [3]

   b) $\frac{2}{3} \div \frac{4}{5}$ [2]

4 A piece of metal is $2\frac{1}{4}$ inches long. Stuart cuts off $\frac{7}{16}$ of an inch. How much is left? [3]

More Exam Practice NE2

Percentages

Finding *A* as a percentage of *B*

- To work out *A* as a percentage of *B*, first write *A* as a fraction of *B*, then change to a decimal and multiply by 100.
- To find an increase or decrease as a percentage, divide the increase or decrease by the original value and multiply by 100.

Chief Examiner Says

Make sure the units are the same.

More Practice N8

Test Yourself (1)

a) Express £30 as a percentage of £80.

b) David's wage increased from £160 to £180 a week. What percentage increase was this?

Chief Examiner Says

If you are allowed to use a calculator, use it, at least to check, even if you feel you can do it in your head.

Percentage increases or decreases

- To find a percentage of an amount, change the percentage to a decimal and multiply.
- To increase or decrease by a percentage, add or subtract the percentage to or from 100. Change this to a decimal (this is the multiplier) and multiply. Alternatively, work out the increase or decrease and add it on or subtract it.

Test Yourself (2)

a) £84 is increased by 7%. Find the new amount.

b) £12 is decreased by 8%. Find the new amount.

More Practice N9

Solutions

Test Yourself (1)

a)
$$\frac{30}{80} = 0.375$$
$$\text{Percentage} = 0.375 \times 100$$
$$= 37.5\%$$

b) Increase = £20
$$\frac{20}{160} = 0.125$$
Divide the increase by the original.
$$\text{Percentage} = 0.125 \times 100$$
$$= 12.5\%$$

Test Yourself (2)

a) 100 + 7 = 107   OR   7% as a decimal is 0.07
Decimal = 1.07 $0.07 \times 84 = 5.88$
New amount   New amount
$= 84 \times 1.07$ $= 84 + 5.88$
$= £89.88$ $= £89.88$

b) 100 − 8 = 92   OR   8% as a decimal is 0.08
Decimal = 0.92 $0.08 \times 12 = 0.96$
New amount   New amount
$= 12 \times 0.92$ $= 12 - 0.96$
$= £11.04$ $= £11.04$

Repeated percentage change

● For repeated change, use the multiplier for each repeat. For example, to increase by 6% five times, multiply by $(1.06)^5$.

Chief Examiner Says

To check how to work out $(1.06)^5$, see 'Powers and roots' in the calculator section.

Test Yourself (1)

a) Sheila invests £6000 at 4% compound interest. How much will the investment be worth after 10 years?

b) A car depreciates in value by 12% per year. It cost £12 500 when new. How much will it be worth after
  i) 2 years?   **ii)** 5 years?

More Practice N10

Chief Examiner Says

Compound interest is repeated percentage increase.

Here is an exam question ...   ... and its solution

a) In a sale, a coat is reduced by £5. The original cost was £40. What percentage reduction was this? **[2]**

b) Sian invested £5500 in a fund. 4% was added to the amount invested at the end of each year. What was the total amount at the end of the 5 years? **[2]**

a) Fraction reduction $= \frac{5}{40}$
$$= 0.125$$
Percentage reduction $= 0.125 \times 100$
$$= 12.5\%$$

b) Total amount $= £5500 \times (1.04)^5$
$$= £6691.59 \text{ (to the nearest penny)}$$

Now Try These Exam Questions

1 Annabel bought a motorcycle for £5595. She later sold it for £4795. Calculate her percentage loss.
Give the answer correct to 2 decimal places. **[3 + 1]**

2 A calculator was sold for £6.95 plus VAT when VAT was 17.5%.
What was the selling price of the calculator including VAT?
Give the answer to the nearest penny. **[3 + 1]**

3 The Candle Theatre has 320 seats.
At one performance 271 seats were occupied.

What percentage of the seats was occupied? Give the answer correct to 2 decimal places. **[2 + 1]**

4 All clothes in a sale were reduced by 15%. Mark bought a coat in the sale that was usually priced at £80.
What was its price in the sale? **[3]**

5 A house went up in value by 1% per month in 2007. At the beginning of the year it was valued at £185 000.
What was its value six months later?
Give the answer to the nearest pound. **[2 + 1]**

More Exam Practice NE3

Solutions

Test Yourself (1)

a) Value $= 6000 \times (1.04)^{10}$
$$= £8881.47 \text{ (to the nearest penny)}$$

b) i) Value after 2 years $= 12\,500 \times 0.88^2$
$$= £9680$$
ii) Value after 5 years $= 12\,500 \times 0.88^5$
$$= £6596.65 \text{ (to the nearest penny)}$$

Ratio

Writing as a ratio

- If two quantities are in proportion to one another so that for x parts of the first there are y parts of the second, then the ratio is x to y, which is written as $x : y$.
- To write a ratio in its simplest form, both parts need to be in the same units and then divided by the same number until they have no common factors.
- To write a ratio in the form $1 : n$, divide the second part by the first.

Test Yourself (1)

Write each of these ratios
i)   in its simplest form.
ii) as $1 : n$.

a) $6 : 9$

b) $50p : £4$

c) $5\,m : 40\,cm$

More Practice N11

Mixing in a ratio

- If two substances are in the ratio $1 : n$
  - to find the second amount from the first, the first must be multiplied by n.
  - to find the first amount from the second, the second must be divided by n.
- If two substances are in the ratio $m : n$
  - to find the second from the first, the first must be multiplied by $\dfrac{n}{m}$.
  - to find the first from the second, the second must be multiplied by $\dfrac{m}{n}$.

Test Yourself (2)

In a recipe, flour and fat are used in the ratio $5 : 2$.

a) How much fat is needed when 480 g of flour is used?

b) How much flour is needed when 300 g of fat is used?

More Practice N12

Solutions

Test Yourself (1)

a) i) $2 : 3$ ii) $1 : 1.5$

> Divide both sides by 3.   Divide both sides by 2.

b) i) $50 : 400 = 1 : 8$ ii) $1 : 8$

> Change to pence and divide both sides by 50.

c) i) $500 : 40 = 25 : 2$ ii) $1 : \frac{2}{25} = 1 : 0.08$

> Change to cm and divide both sides by 20.   Divide both sides by 25.

Test Yourself (2)

a) Fat $= 480 \times \frac{2}{5}$
       $= 192\,g$

b) Flour $= 300 \times \frac{5}{2}$
         $= 750\,g$

Sharing in a given ratio

- To share in a given ratio, first add the parts of the ratio together. Divide the amount to be shared by this total. This is then the multiplier for each of the parts.
- If the amount to be shared is not given, the multiplier can be found by dividing the part given by its part of the ratio. This can then be used to find the other parts or the total.

Chief Examiner Says

If the multiplier is not an exact amount, leave it as a fraction and round the final answer if necessary.

Test Yourself (1)

£150 is shared in the ratio 2 : 3 : 5. How much is each part?

Test Yourself (2)

Exam marks are to be split for Number; Algebra; Shape and Space; and Data Handling in the ratio 6 : 5 : 2 : 3.
There are 42 marks for Number.
How many marks are there all together?

More Practice N13

Direct proportion

- If quantities vary in direct proportion, it means that if you multiply one quantity by a number, you multiply the other by the same number. That is, if you double one quantity, you double the other; if you halve one, you halve the other, and so on.

Test Yourself (3)

The total cost of books is directly proportional to the number of books bought.
If 30 books cost £240, how much does it cost for

a) 120 books?   **b)** 15 books?

More Practice N14

Here is an exam question ...   ... and its solution

John and Peter did some gardening. They shared the money they were paid in the ratio of the number of hours they worked.
John worked for 5 hours. Peter worked for 7 hours. They were paid a total of £28.80.
How much did each one receive?   **[2]**

Ratio is 5 : 7
Total = 12
One share = 28.8 ÷ 12
        = £2.40
John receives 5 × 2.40 = £12
Peter receives 7 × 2.40 = £16.80

Check: £12 + £16.80 = £28.80

Solutions

Test Yourself (1)

Total of parts is 2 + 3 + 5 = 10
Multiplier is 150 ÷ 10 = 15
Parts are 2 × 15 = £30,
3 × 15 = £45, 5 × 15 = £75

Chief Examiner Says

Check that
30 + 45 + 75 = 150

Test Yourself (2)

Multiplier = 42 ÷ 6 = 7

Marks for number ÷ number ratio

Total ratio = 6 + 5 + 2 + 3 = 16
Total marks 16 × 7 = 112

Test Yourself (3)

a) 120 books cost
    4 × 240 = £960

120 = 4 × 30

b) 15 books cost
    £240 ÷ 2 = £120

15 = 30 ÷ 2

Number

8

Now Try These Exam Questions

1 Some of the very first coins were made with 3 parts silver to 7 parts gold.

  a) How much gold should be mixed with 15 g of silver in one of these coins? **[2]**

  b) Another coin made this way has a mass of 20 g.
  How much gold does it contain? **[2]**

2 A recipe for rock cakes uses 100 g of mixed fruit and 250 g of flour. This makes 10 rock cakes.
Jason wants to make 25 rock cakes.
How much mixed fruit and flour does he need? **[2]**

3 A car park contains vans and cars. The ratio of vans to cars is 1 : 6. There are 420 vehicles in the car park.

  a) How many vans are there?

  b) How many cars are there? **[2]**

4 Adrian, Penelope and Gladys shared a lottery win in the ratio 2 : 5 : 8.
They won £7000.
How much did each receive, correct to the nearest penny? **[3 + 1]**

More Exam Practice NE4

Working without a calculator

Multiplying and dividing decimals

● When multiplying decimal numbers, multiply the numbers without the decimal point, then count the total number of decimal places in the question. There must be the same number in the answer.
● When dividing by a number with one decimal place, first multiply both numbers by ten and then carry out the calculation.

Test Yourself (1)

Work out these.
a) 47.3×1.5
b) $4.68 \div 0.6$

More Practice N15

Solutions

Test Yourself (1)

a)
```
    473
×    15
  2365
  4730
  7095
```

There are two decimal places in the question, so there are two in the answer.

So $47.3 \times 1.5 = 70.95$

b) $4.68 \div 0.6 = 46.8 \div 6$

```
      7.8
 6)46.8
  −42.0
    4.8
   −4.8
      0
```

Multiply both numbers by 10.

$4.68 \div 0.6 = 7.8$

Rounding to a given number of significant figures

- In any number, the first digit from the left which is not 0 is the first significant figure.
- When rounding to a given number of significant figures, if the first digit not required is 5 or more, add 1 to the last significant figure and, if necessary, include the correct number of zeros to show the size of the number.

Chief Examiner Says

The number 0.96 correct to 1 significant figure is 1, not 1.0, which has 2 significant figures. Extra zeros are used only to show the size of the number, as in 200, for example.

Test Yourself (1)

Write the following figures correct to
i) 1 significant figure.
ii) 2 significant figures.
a) 3.658 b) 0.0543
c) 935.4 d) 427 510

More Practice N16

Estimating

- To estimate answers to problems, round each number to 1 significant figure and then carry out the calculation without a calculator.

Chief Examiner Says

Round each number to 1 significant figure. If the answer does not work out exactly, give the answer correct to 1 or 2 significant figures. '≈' means 'approximately equals'.

Test Yourself (2)

Estimate the answer to each of these calculations.
a) 47.5×12.6
b) $628 \div 43$
c) $\dfrac{24 \times 6.99}{3.4}$

More Practice N17

Reciprocals

- The reciprocal of x is $\dfrac{1}{x}$.
- The reciprocal of $\dfrac{a}{b}$ is $\dfrac{b}{a}$.

Test Yourself (3)

Find the reciprocal of each of these.
a) 4   b) $\frac{2}{3}$ c) 0.1

More Practice N18

Solutions

Test Yourself (1)

a) i) 4   ii) 3.7
b) i) 0.05   ii) 0.054
c) i) 900   ii) 940
d) i) 400 000 ii) 430 000

Test Yourself (2)

a) $50 \times 10 = 500$
b) $600 \div 40 = 15$
c) $\dfrac{20 \times 7}{3} = \dfrac{140}{3}$
$= 46.666 \ldots$
$\approx 50 \text{ or } 47$

Test Yourself (3)

a) $\frac{1}{4}$ or 0.25
b) $\frac{3}{2}$ or $1\frac{1}{2}$
c) $\frac{1}{0.1}$ or 10

Using place value

- When the solution to a decimal calculation is known, the result of a similar or inverse calculation with the decimal points in different places can be found by multiplying or dividing by an appropriate power of ten.

Test Yourself (1)

Given that $4.7 \times 3.6 = 16.92$,
find the value of each of these calculations.

a) 47×36 **b)** 470×3.6 **c)** $1692 \div 4.7$

More Practice N19

Here is an exam question ...   ... and its solution

a) To change kilograms to pounds, multiply the number of kilograms by 2.2.

Change 4.3 kg to pounds. **[3]**

b) There are 4.546 09 litres in a gallon.
Round 4.546 09

  i) to 3 decimal places. **[1]**
  ii) to 3 significant figures. **[1]**

a)
```
      43
   ×  22
   ─────
      86
     860
   ─────
     946
```
$4.3 \times 2.2 = 9.46$ pounds

b) i) 4.546   **ii)** 4.55

Now Try These Exam Questions

1 The attendance at a world cup match was 54 682. The TV commentator said that there were 55 000 people.
To how many significant figures was the commentator's number? **[1]**

2 Calculate these.
  a) $9.6 - 1.6 \times 2.4$ **[3]**
  b) $7.8 \times (12.9 - 9.4)$ **[3]**
  c) 3.6×1000 **[1]**

3 Petra buys a magazine costing £1.79 every week. Estimate how much she spends on these magazines in a year. **[2]**

4 Write down the value of each of these.
  a) $4.3 \div 100$ **[1]**
  b) The reciprocal of 7 **[1]**
  c) 14.89 correct to 1 significant figure **[1]**

5 Work out $147.2 \div 3.2$. **[3]**

More Exam Practice NE5

Solutions

Test Yourself (1)

a) $47 \times 36 = 1692$

$\times 10, \times 10 = \times 100$

b) $470 \times 3.6 = 1692$

$\times 100$, no change $= \times 100$

c) $1692 \div 4.7 = 360$

$\times 100$, no change $= \times 100$

Calculator methods

Getting to know your calculator

- Calculators vary. The position and symbols used on the buttons differ, depending on the make and model of calculator. The order in which the buttons have to be pressed can also vary. Make sure you know how your calculator works. Don't borrow a different calculator or change your calculator just before an exam. If you are unsure, ask your teacher.

- To input fractions, use the a^b/c button.
  For example, to input $2\frac{3}{4}$, key in 2 a^b/c 3 a^b/c 4 .

Chief Examiner Says

To input -5 on some calculators you need to input 5 $(-)$ rather than $(-)$ 5 . Make sure you know which is needed on your calculator.

Test Yourself (1)

a) Write $\frac{25}{100}$ as a fraction in its simplest form.

b) Work out $1\frac{2}{3} + \frac{3}{4}$.

More Practice N20

Reciprocals

- Use the $1/x$ or the x^{-1} button.

More Practice N21

Test Yourself (2)

Work out the reciprocal of each of these.

a) 0.1   b) 40   c) $1\frac{2}{3}$

Powers and roots

- Use x^y or $\wedge$ for powers.
- Use $\sqrt{\ }$ for square roots and $\sqrt[3]{\ }$ for cube roots.

More Practice N22

Test Yourself (3)

Work out these.

a) 6^7 b) $4^5 + 3^4$

c) $100^{-\frac{1}{2}}$

Solutions

Test Yourself (1)

a) Key in 2 5 a^b/c 1 0 0 $=$

Display $= 1 \lrcorner 4 = \frac{1}{4}$

b) Key in
1 a^b/c 2 a^b/c 3 $+$ 3 a^b/c 4 $=$

Display $= 2 \lrcorner 5 \lrcorner 12$
$= 2\frac{5}{12}$

Test Yourself (2)

a) Key in 0 $.$ 1 $1/x$ $=$ 10

b) Key in 4 0 $1/x$ $=$ 0.025

c) Key in 1 a^b/c 2 a^b/c 3 $1/x$ $=$ $\frac{3}{5}$

Test Yourself (3)

a) Key in 6 x^y 7 $=$ 279 936

b) Key in 4 x^y 5 $+$ 3 x^y 4 $=$ 1105

c) Key in 1 0 0 $x^{1/y}$ $+/-$ 2 $=$ 0.1

or 1 0 0 $\wedge$ $+/-$ 1 a^b/c 2 $=$ 0.1

Depending on your calculator, you may need to press the $+/-$ or $(-)$ button after 2 .

Here is an exam question ...

Work out the following. Give your answers to 2 decimal places.

a) 4.2^4 [1]

b) $\dfrac{3.9^2 + 0.53}{3.9 \times 0.53}$ [2]

c) 350×1.005^{12} [1]

... and its solution

a) 311.17

Key in

4 . 2 x^y 4 =

311.1696

b) 7.61

Key in

(3 . 9 x^2 + 0 . 5 3) ÷
(3 . 9 × 0 . 5 3) =

7.614 900 ...

c) 371.59

Key in

3 5 0 × 1 . 0 0 5 x^y 1 2 =

371.587 234 ...

Now Try These Exam Questions

1 Work out these, giving the answers to 2 decimal places.

   a) 3.4^5 [1]

   b) $(5.1 + 3.7) \times 4.2$ [1]

   c) $\dfrac{5.1 \times 2.6}{14.2 - 6.3}$ [2]

2 Work out the reciprocal of each of these. Give your answers to 2 decimal places where appropriate.

   a) 50 [1]

   b) 0.75 [1]

   c) 3^2 [1]

3 Work out these.

   a) $\frac{3}{5}$ of 200 g [1]

   b) $2\frac{3}{4} - 1\frac{4}{5}$ [2]

   c) $\frac{4}{7}$ of £26.60 [1]

4 Work out these, giving your answers to 2 decimal places where appropriate.

   a) 730×1.01^{15} [1]

   b) $14^{\frac{1}{3}}$ [1]

   c) $\dfrac{840 \times 1.03}{840 + 1.03}$ [2]

More Exam Practice NE6

Solving problems

Some common money problems

You need to be able to
- assess value for money – by working out and comparing cost per unit.
- deal with other ratio problems.
- calculate currency exchange, by multiplying or dividing by an exchange rate.
- calculate tax and insurance.
- work out percentage problems.
- work out gas and electricity bill problems.

Chief Examiner Says

In a question that asks for the best value, if you show no working, you will score no marks, even if you pick the correct item.

Test Yourself (1)

Ali bought a pair of jeans for $38 in the U.S.A.
The exchange rate was $2.03 to £1.
Peter bought a pair of jeans in France for €30.
The exchange rate was €1.44 to £1.
Who paid less for the jeans in £, and by how much?

More Practice N23

Compound measures

These are all examples of compound measures.
- Speed = Distance ÷ Time
- Density = Mass ÷ Volume
- Population density = Population ÷ Area

For more work on these, see the section on Shape and space.

Rounding your answers and checking your work

- Sometimes you are asked to round your answer to a given number of decimal places or a given number of significant figures.
- Sometimes you need to work to a reasonable degree of accuracy. To decide what is reasonable, look at the context and the accuracy of the information you have been given. Your answer cannot be more accurate than the figures you used to calculate it.
- You should always check your work, using at least one of the following methods.
  - Common sense – is the answer reasonable?
  - Inverse operations – work backwards to check.
  - Estimates – one significant figure is usually the best.

Test Yourself (2)

a) Work out £792 ÷ 19.

b) Find an estimate to check that your answer is about right

More Practice N24

Solutions

Test Yourself (1)

Ali paid $38 ÷ 2.03 = £18.72
Peter paid €30 ÷ 1.44 = £20.83
So Ali paid less, by £2.11.

Test Yourself (2)

a) £41.68

Here it is reasonable to give the answer to the nearest penny.

b) 800 ÷ 20 = £40
It checks.

Here is an exam question ...

Tom's Telephone, Internet and Television package costs him

£10.99 a month for the telephone line rental, plus 4.5p per call

£12.99 per month for internet access plus 40p an hour usage

£15.99 for the special television package.

VAT at 17.5% is then added to the total.

One month he made 47 phone calls and used the internet for 35 hours.

How much was his total bill? **[5]**

... and its solution

Telephone cost = £10.99 + 47 × 4.5p
$$= £10.99 + 47 × £0.045$$
$$= £10.99 + £2.115$$
$$= £13.105$$
Internet cost = £12.99 + 35 × 40p
$$= £12.99 + 35 × £0.40$$
$$= £12.99 + £14$$
$$= £26.99$$
Television cost = £15.99
Total before VAT
$$= £13.105 + £26.99 + £15.99$$
$$= £56.085$$
Total including VAT
$$= £56.085 × 1.175$$
$$= £65.899\,875$$
Total bill = £65.90 (to the nearest penny)

Chief Examiner Says

Read the question carefully and break down the problem into steps. Ask yourself
- What do I know?
- What do I have to find?
- What methods can I apply?

Now Try These Exam Questions

1 The table shows the prices of different packs of chocolate bars.

Pack	Size	Price
Standard	500 g	£1.15
Family	750 g	£1.59
Special	1.2 kg	£2.49

Find which pack is the best value for money. You must show clearly how you decide. **[4]**

2 In February 2007 the exchange rate between Pounds and US dollars was £1 = $1.93. Liz changed £500 into US dollars. She went on holiday and spent $784. She changed the rest back into pounds. How much did she receive in pounds? **[4]**

3 a) Mrs Brown took 47 students to London. The train tickets cost £20.25 each. She worked out the cost as £770.75.
Do an estimate to show that this must be the wrong answer. **[2]**

b) The next year, the cost of the train ticket was £21.87. By what percentage had the cost of a ticket increased? **[2]**

4 Coverswift charges £27.50 for five days' holiday insurance and an extra £4.50 for every day after that. How much did they charge for a 14-day holiday? **[3]**

More Exam Practice NE7

Use of symbols

Multiplying brackets by single terms

- When expanding brackets, multiply every term within the brackets by the term immediately in front of the bracket.

Test Yourself (1)

a) Expand $3x(x - 2)$.
b) Expand and simplify $3(2a + 3b) - 2(a - 2b)$.

More Practice A1

Common factors

- Look for every number and letter that is common to every term and write these outside the bracket.
- Write the terms in the bracket needed to give the original expression.

Chief Examiner Says

Always try to take out as big a factor as possible. In part **c)** 2 is a common factor, 4 is a common factor, and a is a common factor. The biggest common factor is $4a$.

Test Yourself (2)

Factorise the following.

a) $3a + 6$
b) $2x^2 - 3xy$
c) $4a - 8ac + 16a^2b$

More Practice A2

Multiplying out two sets of brackets

- To expand expressions like $(a + b)(c + d)$, deal with one pair of brackets at a time.
$$(a + b)(c + d) = a \times (c + d) + b \times (c + d)$$
Now the remaining brackets can be expanded.
$$(a + b)(c + d) = ac + ad + bc + bd$$

Test Yourself (3)

Write without brackets and simplify.

a) $(x + 2)(x - 3)$
b) $(2y - 1)(3y + 4)$

More Practice A3

Solutions

Test Yourself (1)

a) $3x^2 - 6x$
b) $6a + 9b - 2a + 4b$
$= 4a + 13b$

Collect like terms.

Test Yourself (2)

a) $3(a + 2)$
b) $x(2x - 3y)$
c) $4a(1 - 2c + 4ab)$

Note that $4a \times 1 = 4a$ so 1 must be in the bracket.

Test yourself (3)

a) $(x + 2)(x - 3) = x(x - 3) + 2(x - 3)$
$= x^2 - 3x + 2x - 6$
$= x^2 - x - 6$

b) $(2y - 1)(3y + 4) = 2y(3y + 4) - 1(3y + 4)$
$= 6y^2 + 8y - 3y - 4$
$= 6y^2 + 5y - 4$

Indices

- The rules of indices are

$$a^m \times a^n = a^{m+n}$$
$$a^m \div a^n = a^{m-n}$$
$$(a^m)^n = a^{m \times n}$$
$$a^0 = 1$$

- These need to be used to manipulate expressions involving letters and numbers.
- Remember $a^1 = a$.

Test Yourself (1)

Simplify the following.
a) $x^5 \times x^2$
b) $x^6 \div x^3$
c) x^0
d) $(x^2)^3$
e) $2x^2 \times 5x^3$
f) $\dfrac{x^3 \times x^5}{x^2}$

 More Practice A4

Here is an exam question ...

a) Expand the brackets and write as simply as possible $2(3x - 4) - 5(x + 3)$. **[2]**
b) Factorise completely $3a^2 + 6ab$. **[2]**
c) Simplify $2a^4 \times 4a^2$. **[2]**

... and its solution

a) $6x - 8 - 5x - 15 = x - 23$.

> Take care with the signs.
> $-5 \times +3 = -15$

b) $3a(a + 2b)$

> $3a$ is common to both terms.

c) $8a^6$

> Multiply the numbers and add the indices.

Now Try These Exam Questions

1 Expand $5s(s^2 - 2)$. **[2]**

2 Multiply out and simplify
$5(a + 2) - 3(a - 1)$. **[2]**

3 Factorise completely $4x^2 - 2x$. **[2]**

4 Simplify $4a^6 \div 2a^2$. **[2]**

 More Exam Practice AE1

Solutions

Test yourself (1)

a) x^7 b) x^3 c) 1 d) x^6 e) $10x^5$ f) x^6

> Add the powers.

> Subtract the powers.

> Multiply the powers.

> $2 \times 5 = 10$
> Add the powers.

> Add the powers to give x^8 on the top, then $8 - 2 = 6$

Linear equations

Brackets in equations

- If the equation has brackets, multiply out the brackets first.

Test Yourself (1)

Solve $3(x + 4) = 5$

More Practice A5

Unknown quantity on both sides of the equation

- If the unknown quantity is on both sides of the equation, rearrange the equation to collect the numbers on one side and the unknown on the other side.
- It is usually easier to move the unknown to the side where the coefficient (number in front) is positive.

Test Yourself (2)

Solve the following.
$5x + 4 = 2x + 19$

More Practice A6

More than one set of brackets

- If there are more than one set of brackets, multiply all the brackets out first.
- Then simplify both sides before rearranging.

Chief Examiner Says

In a question like **(3)** it is a common error to get the sign in front of the 3 wrong.

Test Yourself (3)

Solve the following.
$5(a + 2) - 3(a + 1) = 8$

More Practice A7

Solutions

Test Yourself (1)

$3x + 12 = 5$
$3x = -7$
$x = -\frac{7}{3}$

Test Yourself (2)

$5x = 2x + 15$
$3x = 15$
$x = 5$

Subtract 4 from both sides.

Subtract 2x from both sides.

Test Yourself (3)

$5a + 10 - 3a - 3 = 8$
$2a + 7 = 8$
$2a = 1$
$a = \frac{1}{2}$

$-3 \times 1 = -3$

Fractions in equations

● If the equation has a fraction in it, multiply by the denominator.

More Practice A8

Test Yourself (1)

Solve the following.

$\frac{2}{3}(x + 2) = 4$

Here is an exam question ...   ... and its solution

Solve the following equations.

a) $2(3 - x) = 1$ **[2]**

b) $\dfrac{5x + 8}{3} = 6$ **[3]**

c) $4(x + 7) = 3(2x - 4)$ **[4]**

a) $6 - 2x = 1$
$-2x = -5$
$x = 2\frac{1}{2}$

b) $\dfrac{5x + 8}{3} = 6$
$5x + 8 = 18$
$5x = 10$
$x = 2$

c) $4(x + 7) = 3(2x - 4)$
$4x + 28 = 6x - 12$
$40 = 2x$
$x = 20$

Now Try These Exam Questions

1 Solve $\dfrac{3m}{4} = 9$.   **[2]**

2 Solve $2(y + 3) = 5y$.   **[3]**

3 Solve $4(x + 2) + 2(3x - 2) = 14$.   **[4]**

More Exam Practice AE2

Formulae

Substituting in formulae

When substituting in formulae, remember
● An expression like ab means $a \times b$.
● Multiplication and division are done before addition and subtraction unless brackets tell you otherwise.
● Expressions like $3r^2$ mean $3 \times r^2$, that is you square r and then multiply by 3.

Chief Examiner Says

When substituting numbers in formulae, take care with the negative numbers.

Don't forget $3a^2$ means $3 \times a^2$

Test Yourself (2)

a) If $C = 6b + 3a^2$, find C when
  i) $b = -3$ and $a = 5$
  ii) $b = 4$ and $a = -2$

b) If $y = \dfrac{3a - 2b}{c}$, find y when
  i) $a = 2$, $b = -5$ and $c = 3$
  ii) $a = \frac{1}{4}$, $b = \frac{3}{4}$ and $c = 2$

More Practice A9

Solutions

Test Yourself (1)

$2(x + 2) = 12$
$2x + 4 = 12$
$2x = 8$
$x = 4$

Multiply both sides by 3.

Test Yourself 2)

a) i) $C = 6 \times -3 + 3 \times 5^2$
    $= -18 + 75 = 57$
ii) $C = 6 \times 4 + 3 \times (-2)^2$
    $= 24 + 12 = 36$

b) i) $y = \dfrac{3 \times 2 - 2 \times (-5)}{3} = \dfrac{6 + 10}{3} = 5\frac{1}{3}$

ii) $y = \dfrac{3 \times \frac{1}{4} - 2 \times \frac{3}{4}}{2} = \dfrac{\frac{3}{4} - \frac{6}{4}}{2} = \dfrac{-\frac{3}{4}}{2} = -\dfrac{3}{8}$

Don't forget that $(-2)^2 = +4$

Writing your own formulae

When writing your own formulae
- Make sure you define your letters carefully, including units.
- Make sure you use brackets if you want addition and subtraction done before multiplication and division.
- Use a fraction line not a ÷ sign for divide.

Test Yourself (1)

Write a formula for the perimeter of a semicircle.

More Practice A10

Rearranging formulae

- To change the subject of a formula, use the equation rule of doing the same to both sides to get the new subject on one side of the formula.

Chief Examiner Says

Remember: always do the same operation to both sides.

Test Yourself (2)

Make a the subject of $v = u + at$

More Practice A11

Here is an exam question ... ... and its solution

The price of a handtool of size S cm is P pence. The formula connecting P and S is $P = 20 + 12S$.

a) Calculate the price of a handtool of size 3 cm. **[2]**

b) Calculate the size of a handtool whose price is 95p. **[2]**

c) Rearrange the formula $P = 20 + 12S$ to express S in terms of P. **[2]**

a) $P = 20 + 12 \times 3$
$= 20 + 36$
$= 56$
The price is 56p

b) $20 + 12S = 95$
$12S = 75$
$S = 75 \div 12$
$= 6.25$
The size is 6.25 cm

c) $P \quad 20 - 12S$
$S = \dfrac{P - 20}{12}$

Now Try These Exam Questions

1 Using $u = 9$, $t = 48$ and $a = -\frac{1}{4}$, work out the value of s from the formula $s = ut + \frac{1}{2}at^2$. **[3]**

2 Use the formula $F = 2(C + 15)$ to find an expression for C in terms of F. **[3]**

3 Rearrange the following to give d in terms of e.
$e = 5d + 3$ **[2]**

More Exam Practice AE3

Solutions

Test Yourself (1)

Let r = the radius in cm and P = the perimeter in cm.
Half the circumference $= \frac{1}{2} \times 2\pi r = \pi r$,
so $P = 2r + \pi r$

Test Yourself (2)

$v - u = at$ — Take u from both sides.

$\dfrac{v - u}{t} = a$ — Divide both sides by t. The fraction line acts as a bracket.

$a = \dfrac{v - u}{t}$ — Rewrite with a on the left.

Inequalities

- The symbols used are
  - $>$ means 'is greater than'
  - $<$ means 'is less than'
  - $\geqslant$ means 'is greater than or equal to'
  - $\leqslant$ means 'is less than or equal to'
- Inequalities are solved using the same rules as for solving equations except when multiplying or dividing by a negative number. Then the inequality sign is reversed. This problem can be avoided by moving the unknown on to the side where it is positive.
- Inequalities can be shown on a number line using the convention.
  - ○——→ when 'equals' is not included
  - ●——→ when 'equals' is included

Test Yourself (1)

a) List the integers for which $-2 \leqslant x < 3$.

b) Solve the following.
  i) $3x - 1 \geqslant 11$
  ii) $5 - x < 2$

Show your solutions on a number line.

More Practice A12

Chief Examiner Says

Don't forget that the solution to an inequality is itself an inequality such as $x \geqslant 4$.
You will lose marks if you simply give a value of x.

Here is an exam question ...   ... and its solution

a) Solve $3x + 4 \leqslant 1$.   **[2]**

b) Show your solution to part **a)** on a number line.   **[1]**

a) $3x \leqslant 1 - 4$
   $3x \leqslant -3$
   $x \leqslant \dfrac{-3}{3}$
   $x \leqslant -1$

b)

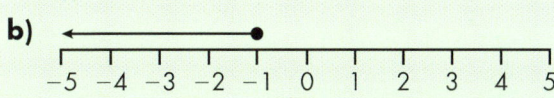

Now Try These Exam Questions

1 List the integers for which $-2 \leqslant x < 5$.   **[2]**

2 Solve these inequalities.
  a) $8x + 5 > 25$ **[2]**

b) $2x + 17 > 4x + 6$ **[2]**
Show the answers on number lines.   **[1][1]**

More Exam Practice AE4

Solutions

Test Yourself (1)

a) $-2, -1, 0, 1, 2$ ◀ 2 is included but 3 is not.

b) i) $3x - 1 \geqslant 11$
   $3x \geqslant 11 + 1$
   $3x \geqslant 12$
   $x \geqslant 4$

'Equals' included so circle is filled in.

ii) $5 - x < 2$ ◀ Add x to both sides so that x is positive.
   $5 < 2 + x$
   $3 < x$ ◀ Subtract 3 from both sides.
   $x > 3$

Swap sides so x is on the left.

'Equals' not included so open circle.

Trial and improvement

- Some equations cannot be solved by algebraic methods. To solve them you need to do a trial and improvement procedure.
- To start trial and improvement you need a first estimate. This could come from a graph, but in examinations you are usually given a hint to the first estimate.
- Start with two values, one which gives too big a result and the other too small. The true solution will be between them.
- Try in between (usually half way).
- Continue the process, finding two values one of which is too big and the other too small until you have an answer to the required accuracy.

Chief Examiner Says

Don't forget to write down the answer to each trial.
It is often best to work in a table.

Here is an exam question

A solution of the equation $x^3 + 4x^2 = 8$ lies between -3 and -3.5. Find this solution by trial and improvement. Give your answer correct to 2 decimal places. **[4]**

... and its solution

$x = -3$
$x^3 + 4x^2 = -27 + 36 = 9$ Too big
$x = -3.5$
$x^3 + 4x^2 = 6.125$ Too small
$x = -3.3$
$x^3 + 4x^2 = 7.623$ Too small
Answer lies between -3.3 and -3
$x = -3.2$
$x^3 + 4x^2 = 8.192$ Too big
Answer lies between -3.3 and -3.2
$x = -3.25$
$x^3 + 4x^2 = 7.921875$ Too small
Answer lies between -3.25 and -3.2

> This solution keeps several decimal places as a check for you. There is no need to write them all down.
> e.g. for $x = 3.23$ $x^3 + 4x^2 = 8.03$ is enough.

$x = -3.23$
$x^3 + 4x^2 = 8.033\,333$ Too big
Answer lies between -3.23 and -3.25
$x = -3.24$
$x^3 + 4x^2 = 7.978\,176$ Too small
Try half way between to check
$x = -3.235$
$x^3 + 4x^2 = 8.005\,89$ Too big
So the answer is between -3.235 and -3.24
and to 2 d.p. the answer is $x = -3.24$.

Solutions

Test Yourself (1)

a)

x	$x^3 + 4x$	Too big / Too small
1	5	Too small
2	16	Too big
1.5	9.375	Too big
1.2	6.528	Too small
1.3	7.397	Too small
1.4	8.344	Too big
1.35	7.860375	Too small

b) (rows 1.2 onward)

These show that x lies between 1 and 2.

So x is between 1 and 1.5.

So x is between 1.3 and 1.4
So correct answer to 1 d.p. is either 1.3 or 1.4
To check which is correct try half way between.

The answer is between 1.35 and 1.4, so $x = 1.4$ correct to one decimal place.

Chief Examiner Says

Don't forget in a question like this that you are trying to find x to 1 decimal place (1.4).
You are not trying to get the value of $x^3 + 4x$ to be 8 to 1 decimal place.

Now Try These Exam Questions

1 The volume of this cuboid is 200 cm³.

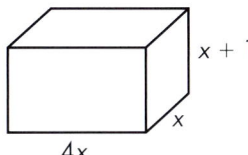

a) Explain why $x^3 + x^2 = 50$. **[2]**

More Exam Practice AE5

b) Find the solution of $x^3 + x^2 = 50$ that lies between 3 and 4.
Give the answer correct to 3 significant figures. You must show your trials. **[4]**

2 a) Show that the equation $x^3 - 8x + 5 = 0$ has a root between $x = 2$ and $x = 3$. **[2]**

b) Use trial and improvement to find this root correct to 1 decimal place.
Show all your trials and their outcomes. **[4]**

Sequences

Common sequences

You should know and recognise the following common sequences.
- Arithmetic sequences, that is those that increase or decrease by a constant amount, for example
  2, 5, 8, 11, 14, ... or 6, 4, 2, 0, −2, ...
- Square numbers: 1, 4, 9, 16, ...
- Powers of 2: 2, 4, 8, 16, ...
- Powers of 10: 10, 100, 1000, 10 000, ...
- Triangle numbers: 1, 3, 6, 10, 15, ...

General rules for sequences

- A formula for the nth term of a sequence makes it possible to obtain any term without finding all the previous terms of the sequence.
- If you have to find a formula for the nth term it will usually be one of the common sequences above.
- Arithmetic sequences which go up for example by 3 each time have an nth term of the form $3n + k$
- Arithmetic sequences which go down for example by 2 each time have an nth term of the form $-2n + k$
- The nth term of the sequence of square numbers is n^2
- The nth term for the sequence of powers of 2 is 2^n
- The nth term for the sequence of powers of 10 is 10^n
- The nth term for the sequence of triangle numbers is $\dfrac{n(n+1)}{2}$
- If you are given a formula for the nth term of a sequence and want to generate the sequence, simply substitute 1 then 2 then 3 then 4 and so on into the formula to find the terms of the sequence.

Test Yourself (1)

a) The nth term of a sequence is $5n - 2$. Find the first four terms of the sequence.

b) Find the nth term of these sequences.
  i) 7, 11, 15, 19, ...
  ii) 10, 7, 4, 1, −2 ...
  Use your answers to find the 50th terms.

More Practice A14

Chief Examiner Says

Don't confuse the term-to-term rule with the rule for the nth term.
A very common mistake is to say that the nth term of a sequence such as 2, 5, 8, 11, ... is $n + 3$ instead of $3n - 1$

Solutions

Test Yourself (1)

a) Substituting 1, 2, 3, 4 gives 3, 8, 13, 18

b) i) The terms increase by 4 each time so the nth term $= 4n + k$
Substituting $n = 1$ gives $4 + k$. This must be 7 so $k = 3$
So nth term $= 4n + 3$. 50th term $= 4 \times 50 + 3 = 203$.

ii) The terms decrease by 3 each time so the nth term $= -3n + k$
Substituting $n = 1$ gives $-3 + k$. This must be 10 so $k = 13$
So nth term $= -3n + 13$ or $13 - 3n$. 50th term $= 13 - 3 \times 50 = -137$.

Chief Examiner Says

Another way of getting k is to imagine that there is a term before the first term.
In part **b) i)** it would be 3 and in part **b) ii)** it would be 13 so the constant terms are 3 and 13.

Here is an exam question ...   ... and its solution

a) Write down the 10th term for the
sequence 3, 7, 11, 15, ... **[1]**

b) Write down an expression for the
*n*th term. **[2]**

c) Show that 137 cannot be a term in this
sequence. **[3]**

a) 10th term = $3 + 9 \times 4 = 39$

b) Either:
the difference between terms is 4 so the
expression will start 4*n*.
If $n = 1$ then $4n = 4$
subtract 1 to get 3
therefore the expression is $4n - 1$.
Or:
1st term is 3, add 4 $(n - 1)$ times therefore *n*th
term is $3 + 4(n - 1) = 3 + 4n - 4 = 4n - 1$.

c) If 137 is in the sequence then $4n - 1 = 137$
$4n = 138$
$n = 138 \div 4 = 34.5$
i.e. not a whole number.
Therefore 137 cannot be in the sequence.

Now Try These Exam Questions

1 The first four terms of a sequence are 2, 9, 16, 23.
   a) Find the rule for this sequence. **[2]**
   b) Show that 300 is not in this sequence. **[3]**

2 These classroom tables seat four children.

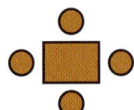   

1 table   2 tables
4 children   8 children

   a) For this arrangement, write down a formula linking the number of children, *c*,
   to the number of tables, *t*. **[1]**

   b) Sometimes the tables are arranged like this:

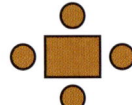   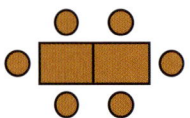   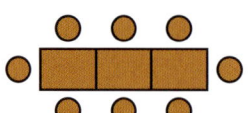

   For this arrangement, write down a formula linking the number of children, *c*,
   to the number of tables, *t*. **[2]**

More Exam Practice AE6

Plotting graphs

Straight lines

- Equations of straight lines are usually given in one of two forms:
  $y = 2x + 1$ or $3x + 2y = 6$
- As with all graphs, the first step is to make a table of the x values and y values.
- With the $y = 2x + 1$ type, choose about three simple values of x
- With the $3x + 2y = 6$ type, chose $x = 0$ and $y = 0$. You may wish to choose another value of x as a check.
- Next plot the coordinate pairs.
- Finally join the points up with a straight line and extend it to cover the grid.

Test Yourself (1)

Draw graphs of the following

a) $y = 2x + 1$

b) $3x + 2y = 6$

More Practice A15

Quadratic graphs

- The equations of quadratic graphs are in the form
  $y = ax^2 + bx + c$
- The shape of quadratic graphs is

    for $a > 0$ for $a < 0$

- The name of this shape graph is a parabola. All parabolas are symmetrical. This symmetry can be seen on the graph and can often be seen in the table as well.
- As with all graphs, the first step is to draw up a table of values. For simple equations you may be able to simply have a row for x and a row for y. For more difficult equations you may wish to put in extra rows, between the x row and the y row, for the steps in the calculation.
- Then plot the points.
- Then join the points with as smooth a curve as possible.

Test Yourself (2)

Draw the graph of
$y = x^2 - 4x + 3$
for $-1 \leqslant x \leqslant 5$.

Chief Examiner Says

Don't forget to join the points. You will lose marks if you do not do so. Also it may be impossible to answer any questions about the graph if you do not draw the curve.

More Practice A16

Solutions

Test Yourself (1)

a)
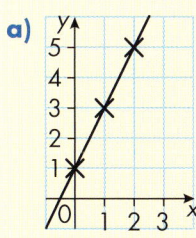

$x = 2, y = 2 \times 2 + 1 = 5$

x	0	1	2
y	1	3	5

b)
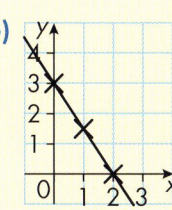

For $x = 0$, $2y = 6$ so $y = 3$

x	0	1	2
y	3	$1\frac{1}{2}$	0

For $x = 1$, $3 + 2y = 6$ so $y = 1\frac{1}{2}$

Test yourself (2)

x	-1	0	1	2	3	4	5
y	8	3	0	-1	0	3	8

and

For example
when $x = -1$
$y = (-1)^2 - 4(-1) + 3$
$= 1 + 4 + 3 = 8$

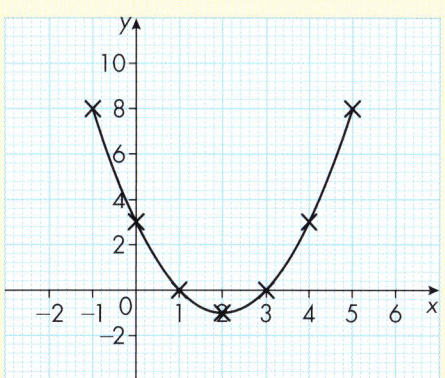

Using graphs to solve quadratic equations

- The solution to $ax^2 + bx + c = k$ is the values of x where the graph $ax^2 + bx + c$ crosses the line $y = k$
- The first step is to plot the graph of $y = ax^2 + bx + c$
- Then draw the line $y = k$
- Then read off the x coordinates where the line crosses the curve. Most quadratic equations have two possible solutions.
- If you are solving $y = ax^2 + bx + c = 0$, read off the x values where the curve crosses the x-axis ($y = 0$)

Chief Examiner Says

Don't forget the solutions to $ax^2 + bx + c = k$ are x values *not* y values. You may lose marks if you include the y coordinates of the points where the curve crosses the line.

Test Yourself (1)

Use the graph of $y = x^2 - 4x + 3$ to solve these equations.

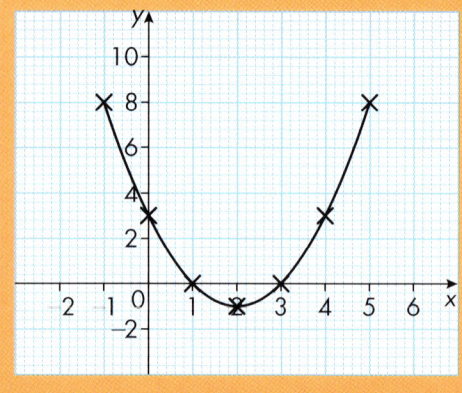

a) $x^2 - 4x + 3 = 0$
b) $x^2 - 4x + 3 = 6$

More Practice A17

Here is an exam question ...   ... and its solution

a) Copy and complete the table for $y = 2x^2 - 3x$ [2]

x	−1	0	1	2	3
y			−1	2	

b) Draw the graph of $y = 2x^2 - 3x$.   [2]
c) Use your graph to solve $2x^2 - 3x = 0$.   [2]

a)

x	−1	0	1	2	3
y	5	0	−1	2	9

b)

c) Solutions where the curve crosses the x-axis:
$x = 0$ or 1.5

Solutions

Test Yourself (1)

a) Look where the graph crosses the x-axis ($y = 0$): the solutions are $x = 1$ and $x = 3$.
b) Look where the graph crosses the line $y = 6$: the solutions are $x = -0.6$ and 4.6 correct to 1 d.p.

Now Try These Exam Questions

1 a) Copy and complete the table of values and draw the graph of $y = x^2 - 2x + 1$ for values of x from -1 to 3. **[4]**

x	−1	0	1	2	3
y		1			4

b) Use your graph to find the values of x when $y = 3$. **[2]**

2 a) Copy and complete the table for $y = 4x - x^2$ and draw the graph. **[4]**

x	−1	0	1	2	3	4	5
y				3		0	

b) Use your graph to find
  i) the value of x when $4x - x^2$ is as large as possible. **[2]**
  ii) between which values of x the value of $4x - x^2 - 2$ is larger than 0. **[2]**

More Exam Practice AE7

Interpreting graphs

Graphs in real situations

To interpret real-life graphs
- Look at the labels on the axes – they tell you what the graph is about
- For distance–time graphs, the steeper the line the greater the speed
- If the graph is a straight line the quantities are increasing or decreasing at a steady rate. If the graph is a curve the rate is not constant
- You can find speed from the relationship: speed = distance ÷ time
- A horizontal line indicates that the quantity on the y-axis is constant. For a distance–time graph this means the object is standing still

Test Yourself (1)

Asif walked to the bus stop, waited for the bus, then travelled on to school. The graph is a distance–time graph for Asif's journey.

a) How long did Asif wait at the bus stop?

b) How far did Asif travel
  i) on foot?   **ii)** by bus?

c) How fast did Asif walk?

d) What was the average speed of the bus?

e) Why are the sections of the graph, in reality, unlikely to be straight?

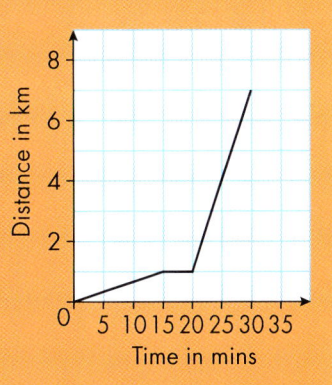

More Practice A18

Solutions

Test Yourself (1)

a) 5 mins

b) i) 1 km   **ii)** 6 km

c) $\frac{1}{15}$ km/min or 4 km/h

d) $\frac{6}{10}$ km/min or 36 km/h

e) Stops and starts are likely to be gradual, not sudden. Speed unlikely to be constant.

Chief Examiner Says

Remember to find the rate of change for each part of the graph. If it is zero, say how long this lasts.

Here is an exam question ...

Tom leaves home at 8.20 a.m. and goes to school on a moped. The graph shows his distance from the school in kilometres.

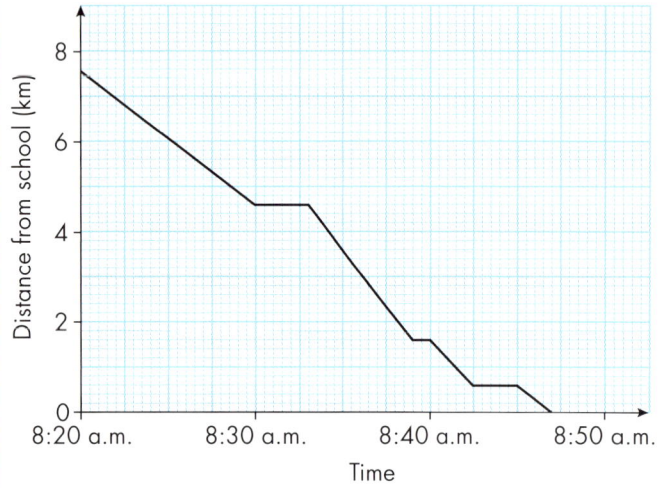

a) How far does Tom live from school? **[1]**

b) Write down the time that Tom arrives at the school. **[1]**

c) Tom stopped three times on the journey. For how many minutes was he at the last stop? **[1]**

d) Calculate his speed in km/h between 8.20 a.m. and 8.30 a.m. **[2]**

... and its solution

a) 7.6 km

b) 8.47 a.m.

c) 2.5 minutes

d) Distance = 7.6 − 4.6 = 3 km
Time = 10 mins
Speed = $\frac{3}{10} \times 60 = 18$ km/h

Now Try These Exam Questions

1 Jim went out walking.
In the diagram ABCD represents his walk.

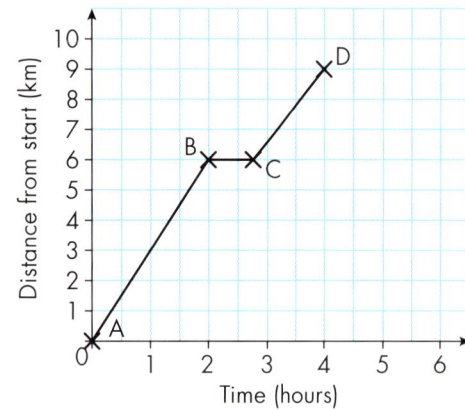

a) How far had Jim walked after $1\frac{1}{2}$ hours? **[1]**

b) What does the part of the graph BC represent? **[1]**

c) After walking 9 km, Jim turned round and walked straight back to his starting place without stopping. It took him 2 hours to get back.
Draw a line on a copy of the grid to show this. **[2]**

d) Work out his average speed on the return journey. **[2]**

More Exam Practice AE8

Angles and two-dimensional shapes

Basic angle facts

You need to know and be able to use these angle facts:
- Angles round a point add to 360°.
- Vertically opposite angles are equal.
- Angles on a straight line add to 180°.
- The angles in a triangle add to 180°.

The last two of these facts can be combined to show:
- The exterior angle of any triangle is equal to the sum of the opposite interior angles.

In the diagram, d is the exterior angle. So $d = b + c$.

Chief Examiner Says

When you give reasons, make sure you give the relevant geometrical reason, not just a calculation. For example, write 'the angles in a triangle add to 180°' not just '$x + 60 + 48 = 180$'.

Test Yourself (1)

Work out the angles marked with letters in these diagrams. Give reasons for your answers.

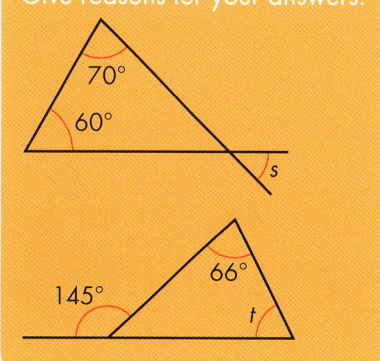

More Practice S1

Angles with parallel lines

- a and b are called alternate angles. They are on opposite sides of the transversal.

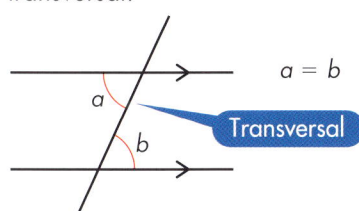

$a = b$

Transversal

- c and d are called corresponding angles. They are in the same position between the transversal and the parallel lines.

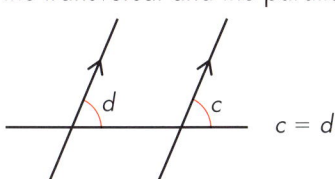

$c = d$

Test Yourself (2)

Work out the angles marked with letters in these diagrams. Give reasons for your answers.

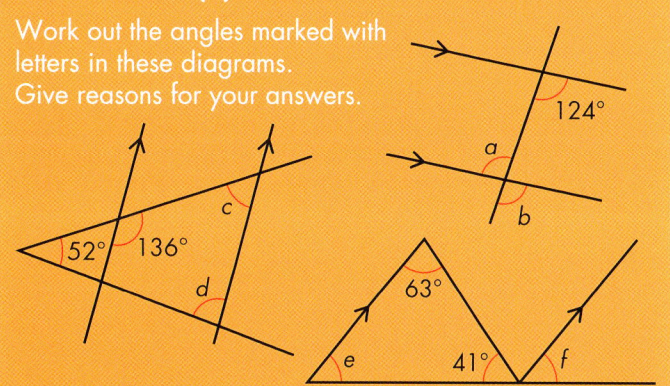

More Practice S2

- e and f are called allied angles. They are the same side of the transversal between the parallel lines.

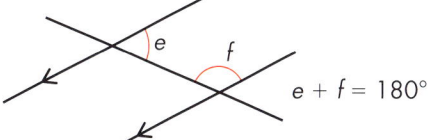

$e + f = 180°$

Solutions

Test Yourself (1)

$s = 50°$ Angles in a triangle add to 180° *and* Vertically opposite angles are equal.

$t = 79°$ Exterior angle of triangle equals sum of interior opposite angles.
(or angles on a straight line add to 180° *and* angles in a triangle add to 180°.)

Test Yourself (2)

$a = 124°$ Alternate angles are equal.

$b = 124°$ Corresponding angles are equal *or* Vertically opposite angles are equal.

$c = 44°$ Allied angles add to 180°.

$d = 84°$ Angles in a triangle add up to 180°.

$e = 76°$ Angles in a triangle add up to 180°.

$f = 76°$ Corresponding angles are equal.

Quadrilaterals

These are the different quadrilaterals and the facts you need to know.

- Square
  - All angles 90°.
  - All sides equal.
  - Opposite sides parallel.
  - Diagonals equal and bisect at 90°.
  - Four lines of symmetry.
  - Rotation symmetry order 4.

- Rectangle
  - All angles 90°.
  - Opposite sides equal and parallel.
  - Diagonals equal and bisect but not at 90°.
  - Two lines of symmetry.
  - Rotation symmetry order 2.

- Parallelogram
  - Opposite angles equal.
  - Opposite sides equal and parallel.
  - Diagonals not equal but bisect, though not at 90°.
  - No line symmetry.
  - Rotation symmetry order 2.

- Rhombus
  - Opposite angles equal.
  - All four sides equal.
  - Opposite sides parallel.
  - Diagonals not equal length but bisect at 90°.
  - Two lines of symmetry.
  - Rotation symmetry order 2.

- Trapezium
  - One pair of opposite sides parallel.

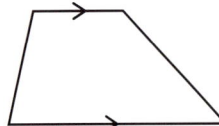

- Isosceles trapezium
  - Two pairs of adjacent angles equal.
  - One pair of opposite sides equal.
  - Other pair of sides parallel.
  - Diagonals equal but do not bisect or cross at 90°.
  - One line of symmetry.
  - No rotation symmetry.

- Kite
  - One pair of opposite angles equal.
  - Two pairs of adjacent sides equal but none parallel.
  - One diagonal bisected at 90°.
  - One line of symmetry.
  - No rotation symmetry.

Chief Examiner Says

Remember what each shape is called and what it looks like. You can then use a sketch and symmetry to help you see which angles are equal, and so on.

Test Yourself (1)

A quadrilateral has both pairs of opposite sides parallel.
What type of quadrilateral can it be?

Test Yourself (2)

Draw two quadrilaterals which have only one line of symmetry.

More Practice S3

Solutions

Test Yourself (1)

It can be a square, a rectangle, a parallelogram or a rhombus.

Test Yourself (2)

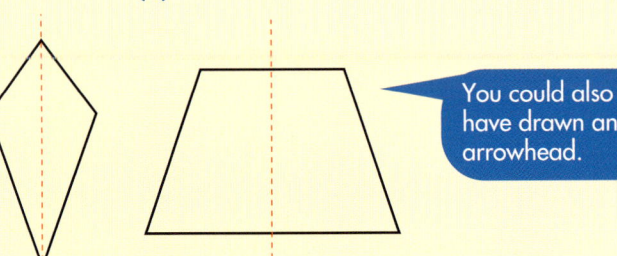

You could also have drawn an arrowhead.

Polygons

- The sum of the interior angles of a quadrilateral is 360°.
- The sum of the interior angles of a pentagon (five sides) is 540°.
- The sum of the interior angles of a hexagon (six sides) is 720°.
- At any vertex of a polygon, the interior and exterior angles make a straight line and add up to 180°.
- The sum of the exterior angles of any polygon is 360°.
- A regular polygon has all its sides equal and all its angles equal.

More Practice S4

Test Yourself (1)

A regular polygon has 12 sides.

a) Work out the size of each exterior angle.

b) Work out the size of each interior angle.

c) Find the sum of the interior angles.

Pythagoras' theorem

- In a right-angled triangle, labelled like this, Pythagoras' theorem states that $a^2 = b^2 + c^2$.
- The longest side of a right-angled triangle is called the hypotenuse.

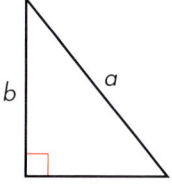

Chief Examiner Says

It is a good idea to label the triangles, using letters for the sides, and to write down the rule; then substitute the actual values for the sides before rearranging, if that is needed.

Chief Examiner Says

As a check, remember that in a right-angled triangle the longest side, the hypotenuse, is always opposite the right-angle.

Test Yourself (2)

Calculate the length of the unknown side in each of these triangles.

a)

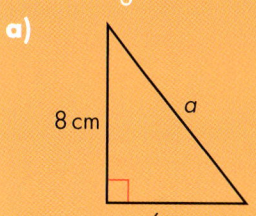

b)

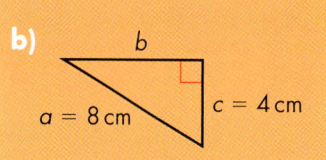

More Practice S5

Solutions

Test Yourself (1)

a) Each exterior angle $= \frac{360}{12}$
$= 30°$

b) Each interior angle $= 180° - 30°$
$= 150°$

c) Sum of the interior angles $- 150° \times 12$
$= 1800°$

Test Yourself (2)

a) $a^2 = b^2 + c^2$ *a is the hypotenuse*
$= 8^2 + 6^2$
$= 64 + 36$
$= 100$
$a = \sqrt{100}$
$= 10 \text{ cm}$

b) $a^2 = b^2 + c^2$
$8^2 = b^2 + 4^2$
$64 = b^2 + 16$
$b^2 = 64 - 16$
$b^2 = 48$
$b = \sqrt{48}$
$b = 6.9 \text{ cm (to 1 d.p.)}$

Areas and perimeters

You should learn these formulae.

- Area of a rectangle
  $A = b \times h$

- Area of a triangle
  $A = \dfrac{b \times h}{2}$

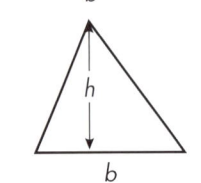

- Area of a parallelogram
  $A = b \times h$

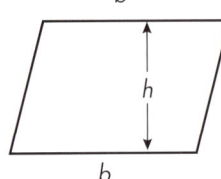

- Area of a trapezium
  $A = \dfrac{h \times (a + b)}{2}$

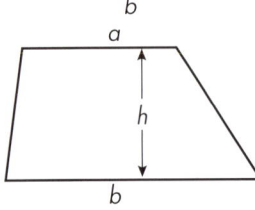

- Area of a circle
  $A = \pi r^2$

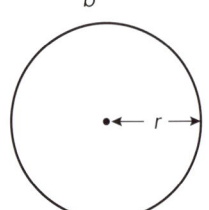

- Circumference of a circle
  $C = \pi d$

> 'Circumference' is just another name for the perimeter of a circle.

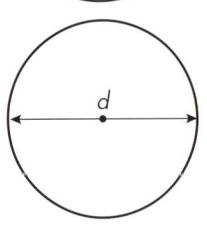

Test Yourself (1)

The vertices of a triangle are at $(2, 1)$, $(2, -3)$ and $(14, -3)$. Draw axes and plot the triangle. Find its area.

Test Yourself (2)

This shape is made from a parallelogram and a semicircle.

a) Find its area.

b) Find its perimeter.

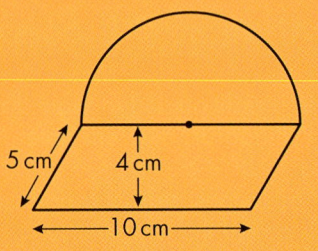

5 cm   4 cm

10 cm

Test Yourself (3)

A circular pond has an area of 40 m². Calculate the radius of the pond.

(More Practice S6)

Solutions

Test Yourself (1)

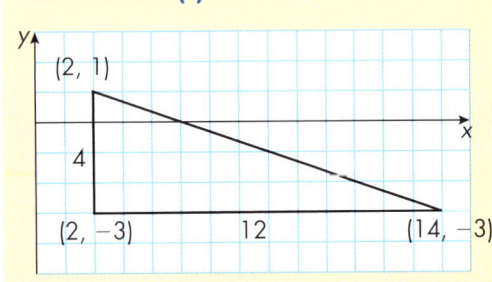

$A = \dfrac{12 \times 4}{2} = 24$ units²

Test Yourself (2)

a) Diameter of circle = 10 cm, so radius = 5 cm
Area of semicircle
$= \tfrac{1}{2} \times \pi \times 5^2$
$= 39.269 \ldots$ cm²
Area of parallelogram
$= 10 \times 4$
$= 40$ cm²
Area of whole shape
$= 79.3$ cm² (to 1 d.p.)

b) Perimeter of shape
$= 5 + 10 + 5 + \tfrac{1}{2} \times \pi \times 10$
$= 35.7$ cm (to 1 d.p.)

Test Yourself (3)

$A = \pi r^2$
$40 = \pi \times r^2$
$\dfrac{40}{\pi} = r^2$
$r = \sqrt{\dfrac{40}{\pi}}$
$r = 3.6$ cm (to 1 d.p.)

Here is an exam question ...

a) The sketch shows a regular pentagon and a regular hexagon with equal length sides.

  i) Work out the size of one interior angle of the pentagon and the hexagon. **[2]**

  ii) Work out the sizes of the angles in triangle ABC. **[2]**

b) A heart shape is made from a square and two semi-circles. Find the area and perimeter of the heart shape. **[3]**

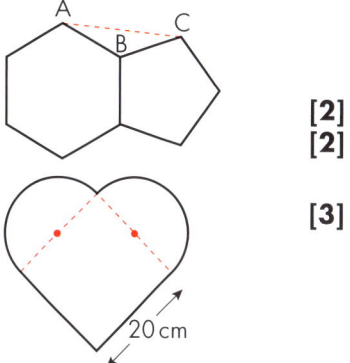

... and its solution

a) i) Exterior angle of pentagon $= \dfrac{360°}{5} = 72°$ Exterior angles of a polygon add up to 360°.

Interior angle $= 180° - 72° = 108°$

Exterior angle of hexagon $= \dfrac{360°}{6} = 60°$

Interior angle $= 180° - 60° = 120°$

ii) Angle ABC $= 360° - 120° - 108° = 132°$ Angles round a point add up to 360°.
Triangle ABC is isosceles.   AB = BC

So angle BAC = angle BCA $= \dfrac{180° - 132°}{2}$ Angles in a triangle add up to 180°.

$= \dfrac{48°}{2} = 24°$

b) Shape = square of side 20 cm + one whole circle of radius 10 cm
Area of shape $= 20 \times 20 + \pi \times 10^2 = 714.2 \text{ cm}^2$ (to 1 d.p.)
Perimeter of shape = two semicircles + two sides of square
$=$ circumference of whole circle $+ 40$ cm
$= \pi \times 20 + 40$
$= 102.8$ cm (to 1 d.p.)

Now Try These Exam Questions

1 In the diagram, BC is parallel to DE.

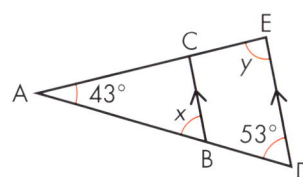

Find the size of x and y. Give reasons for your answers. **[4]**

2 a) Sketch a rhombus and mark everything that is equal.

  b) Draw in all the lines of symmetry.

  c) State the order of rotation symmetry. **[4]**

3 A pentagon has four interior angles of 75°, 96°, 125° and 142°. Find the size of the fifth interior angle. **[3]**

4 a) Find the area of this triangle.

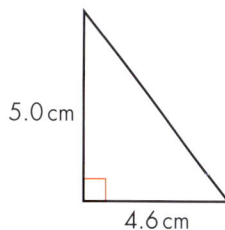

  b) Calculate the length of the hypotenuse of this triangle. Give your answer to a sensible degree of accuracy. **[5]**

More Exam Practice SE1

Three-dimensional shapes

The net of a solid

- There is more than one way of drawing the net of a 3D shape. For example, here are some possible nets for a cube.

Chief Examiner Says

For the net of a pyramid or triangular prism, for example, you may also need to use the skills of constructing a triangle. See page 43.

Test Yourself (1)

Draw accurately on squared paper the net of this cuboid.

1 cm   3 cm
2 cm

More Practice S7

Isometric drawing

- 3D shapes can be represented on isometric paper.

Chief Examiner Says

Right angles appear as 60° or 120° when drawn on isometric paper.

Test Yourself (2)

Make a drawing on isometric paper of a cuboid measuring 3 cm by 4 cm by 2 cm.

More Practice S8

Plans and elevations

- The plan view of an object is the view seen directly from above.
- The front elevation is the view seen directly from the front.
- The side elevation is the view seen directly from one side (there are two possible side elevations).

Test Yourself (3)

Sketch the plan (P), the front elevation (F) and side elevations (S₁ and S₂) of this shape.

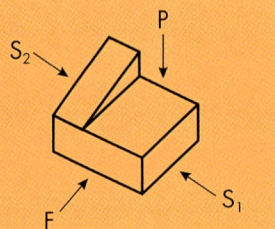

Chief Examiner Says

Remember that hidden edges are shown as dotted lines and right angles should be right angles.

More Practice S9

Solutions

Test Yourself (1)

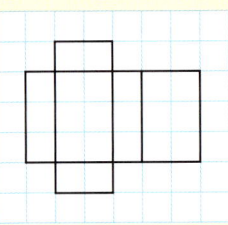

Not full size.

Test Yourself (2)

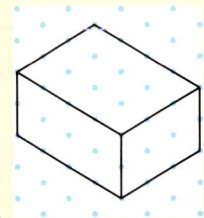

Test Yourself (3)

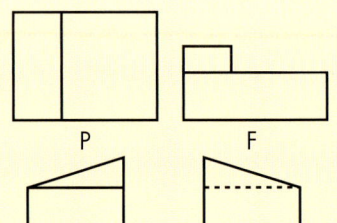

Volume and surface area of prisms

- A prism is a three-dimensional shape with a constant cross-section.
- These are some examples of prisms.

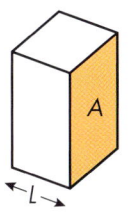
Cuboid

Cylinder

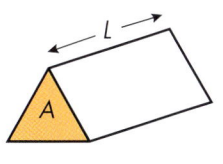
Triangular prism

Chief Examiner Says

Don't forget to include the units in your answer.

- For all prisms, Volume = Area of cross-section × length, or $V = A \times L$.
- The surface area of a prism is the total area of all the surfaces of the prism.

Chief Examiner Says

Don't use rounded answers in the middle of a calculation. Keep all the figures on your calculator (e.g. use the answer function) and just round at the end. Give your answer to a sensible degree of accuracy if no specific accuracy is stated.

Test Yourself (1)

a) Find the volume and surface area of this cuboid.

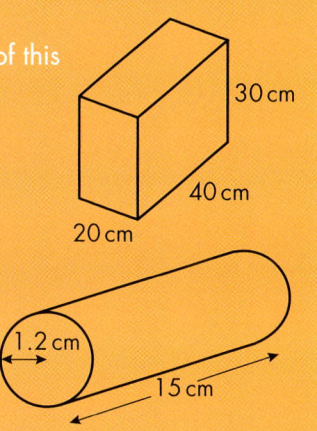

30 cm
40 cm
20 cm

b) Find the volume of this packet of sweets.

1.2 cm
15 cm

More Practice S10

Solutions

Test Yourself (1)

a) Volume of cuboid = length × width × height
$$= 40 \times 20 \times 30$$
$$= 24\,000 \text{ cm}^3$$

Surface area = 2 × top + 2 × side + 2 × front
$$= 2(20 \times 40) + 2(40 \times 30) +$$
$$2(20 \times 30)$$
$$= 1600 + 2400 + 1200$$
$$= 5200 \text{ cm}^2$$

b) Area of cross-section $= \pi \times 1.2^2$
$$= 4.52... \text{ cm}^2$$

Volume = Area of cross-section × length
$$= 4.52... \times 15$$
$$= 67.9 \text{ cm}^3 \text{ (to 1 d.p.)}$$

Here is an exam question ...   ... and its solution

Find the volume of this greenhouse.
The ends are semi-circles.

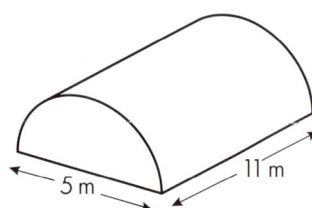

5 m   11 m

[3]

Area of end = $\frac{1}{2} \times \pi r^2$

$\qquad = \frac{1}{2} \times \pi \times 2.5^2$

Volume = area of end × length

$\qquad = (\frac{1}{2} \times \pi \times 2.5^2) \times 11$

$\qquad = 108\ m^3$ (to 3 s.f.)

Now Try These Exam Questions

1 This sweet box is in the shape of a prism. The base is an isosceles right-angled triangle.

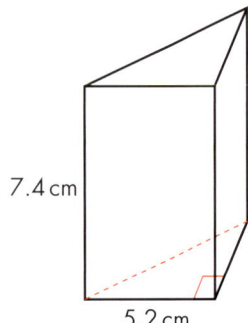

7.4 cm

5.2 cm

Construct the net of the box.   **[4]**

2 a) How many faces does this L-shaped prism have?

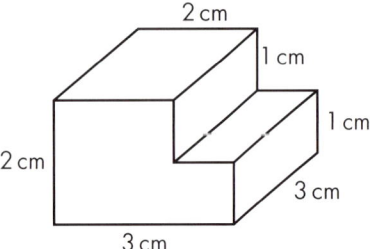

2 cm

1 cm

1 cm

2 cm

3 cm

3 cm

b) How many vertices does it have?

c) Make an isometric drawing of this prism.

d) Calculate the volume of this prism.   **[6]**

3 Sketch the plan (P) and side elevation (S) of this shape.

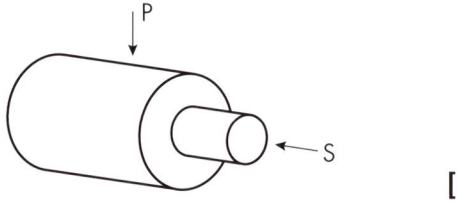

P

S

[3]

4 This is a triangular prism.

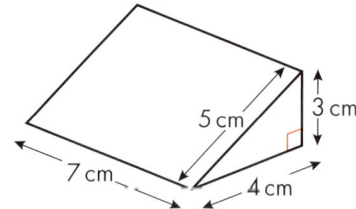

5 cm   3 cm

7 cm   4 cm

a) Find its volume.

b) Find its surface area.   **[6]**

More Exam Practice SE2

Transformations and coordinates

Chief Examiner Says

At GCSE, you will meet four types of transformations: reflection, rotation, enlargement and translation.

You may be asked to follow instructions and draw the result of a transformation or to describe a transformation that has taken place. When describing a transformation, give the type of transformation first, then the extra information required.

Reflection

- The image is the same shape and size as the object, but is reversed.
- The distance between each point on the image and the mirror line is the same as the distance between the mirror line and the corresponding point on the object.
- When describing a reflection, state the mirror line.

Test Yourself (1)

Reflect this triangle in the line $x = 3$.

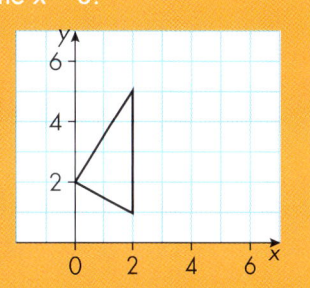

More Practice S11

Rotation

- The image is the same shape and size as the object, but is turned round.
- The distance between each point on the image and the centre of rotation is the same as the distance between the centre of rotation and the corresponding point on the object.
- Use tracing paper and trial and improvement to help you find the centre of rotation.
- When describing a rotation, state
  - the centre of rotation.
  - the angle of rotation.
  - the direction – clockwise or anticlockwise.

Test Yourself (2)

Rotate this flag through 90° clockwise about (0, 2).

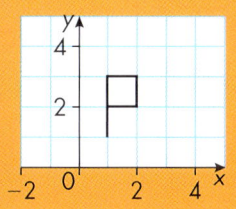

More Practice S12

Solutions

Test Yourself (1)

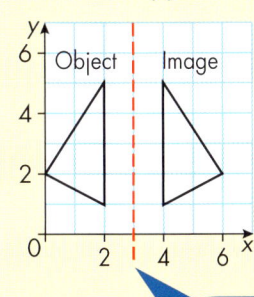

The line $x = 3$ is the vertical line through 3 on the x-axis.

Test Yourself (2)

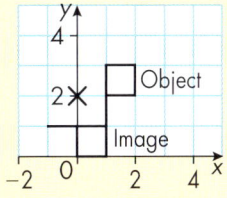

Enlargement

- The image is the same shape as the object but each length on the image is the corresponding length on the object multiplied by the scale factor. This is the only transformation you meet at GCSE where the object and the image are not congruent. When two things are the same shape but a different size like this, they are called similar shapes.
- The distance of each point on the object from the centre of enlargement is multiplied by the scale factor to find the distance from the centre to the corresponding points on the image.
- To find the centre of enlargement, join corresponding points on the object and image and extend the lines until they meet.
- When describing an enlargement, state
  - the centre of enlargement.
  - the scale factor.

Test Yourself (1)

Enlarge this flag with centre (0, 3) and scale factor 2.

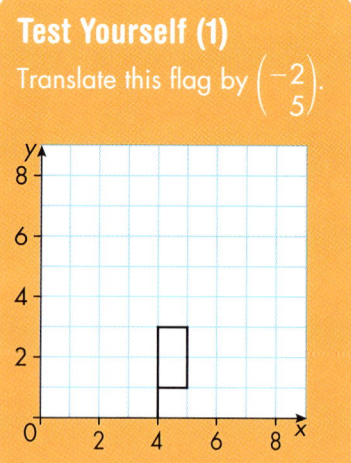

More Practice S13

Translations

- Each point on the image has moved the same distance, in the same direction, from the corresponding point on the object.
- The image is the same size and shape as the object and is the same way up.
- When describing a translation, state the column vector, or how many units the shape has moved in each direction.

Chief Examiner Says

In an examination, when asked to describe a single transformation, you have first to work out which type of transformation is needed. If the object and image are not the same size, check that it is an enlargement. If they are the same size, draw the object on tracing paper and move the tracing paper until it fits on the image. If you just move the tracing paper in a straight line, it is a translation. If you need to turn the tracing paper round, it is a rotation. If you turn the tracing paper over, it is a reflection.

Test Yourself (1)

Translate this flag by $\begin{pmatrix} -2 \\ 5 \end{pmatrix}$.

More Practice S14

Solutions

Test Yourself (1)

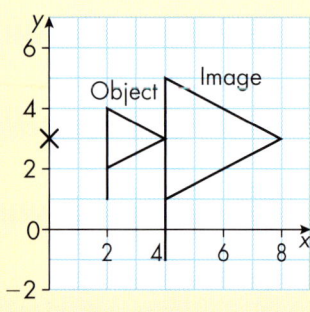

Test Yourself (2)

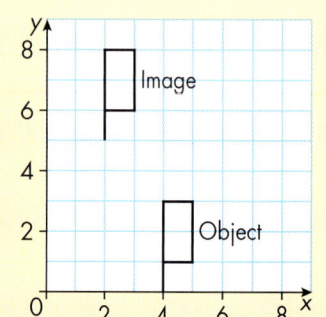

Translate by the vector $\begin{pmatrix} -2 \\ 5 \end{pmatrix}$ means move 2 units across to the left and 5 units up.

Coordinates

- You may be asked to complete a diagram, for instance to complete a square and give the coordinates of the missing corner.
- To find the midpoint of the line joining A and B, find the mean of their coordinates.
- To describe a position in three dimensions, where there is height too, there are three coordinates.

Test Yourself (1)

Calculate the midpoint of AB when A is (1, 5) and B is (7, 2).

Test Yourself (2)

ABCDEFGH is a cuboid. Write down the coordinates of the following points.

a) C **b)** H **c)** F

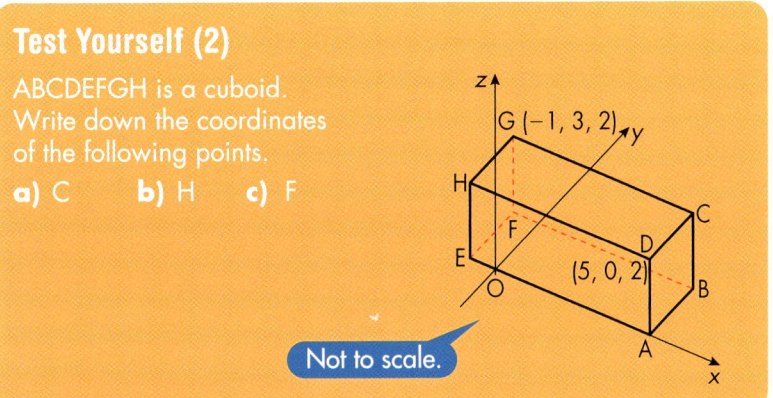

Not to scale.

More Practice S15

Here is an exam question ...

a) Triangle T is rotated 180° clockwise about the point (0, 0). Its image is triangle R. Draw and label triangle R. **[2]**

b) Triangle R is reflected in the *y*-axis. Its image is triangle S. Draw and label triangle S. **[1]**

c) Describe the single transformation which would map triangle T on to triangle S.

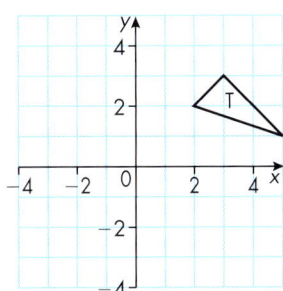

[3]

... and its solution

a) and **b)**

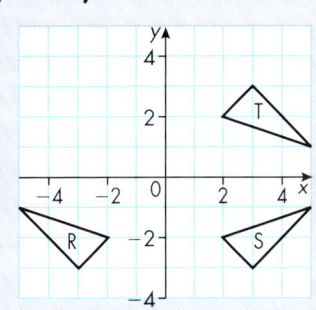

c) Reflection in the *x*-axis.

Solutions

Test Yourself (1)

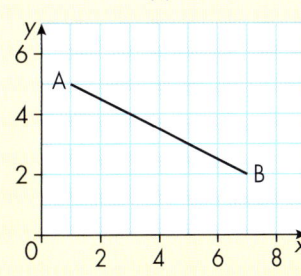

The midpoint of AB is $\left(\dfrac{1+7}{2}, \dfrac{5+2}{2}\right) = (4, 3.5)$

Test Yourself (2)

a) (5, 3, 2)
b) (−1, 0, 2)
c) (−1, 3, 0)

Now Try These Exam Questions

1 The diagram shows shapes A and B.

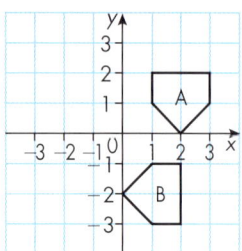

a) Describe fully the single transformation that maps shape A on to shape B.

b) Draw the image of shape A after a reflection in the *y*-axis. Label the image C.

c) Draw the shape A after an enlargement with centre (0, 0) and scale factor 3. Label the image D. Note that you will need an *x*-axis from −5 to 10 and a *y*-axis from −5 to 8.

[7]

2 The diagram shows a cuboid.

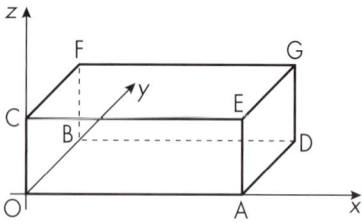

A is the point (6, 0, 0).
D is the point (6, 4, 0).
C is the point (0, 0, 2).

Write down the coordinates of the following points.

a) B

b) F

c) G

d) the midpoint of AD **[4]**

More Exam Practice SE3

Measures

Bearings and scale drawings

- Bearings are measured clockwise from North and must have three figures. An angle of 27° would be written as a bearing of 027°.
- The bearing of B from A is the angle at A, clockwise from North to the line to B, marked *x* in the diagram. The bearing of A from B is the angle marked *y* in the diagram

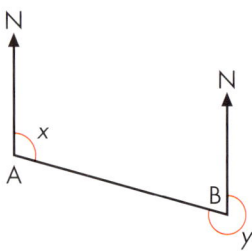

- Scales are sometimes expressed as ratios. For example, a scale of 1 : 10 000 on a map means that 1 cm on the map represents 10 000 cm (which is 100 m) on the ground.

Test Yourself (1)

Measure the bearings of A, B and C from O.

More Practice S16

Chief Examiner Says

A useful rule to remember is that the difference between the bearing of B from A and the bearing of A from B is 180°.

Solutions

Test Yourself (1)

A 050°
B 140°
C 305°

Changing units

- As well as changing between linear metric units, you need to know how to change area and volume units.

  Length: $1 \text{ cm} = 10 \text{ mm}$ $1 \text{ m} = 100 \text{ cm}$

  Area: $1 \text{ cm}^2 = 10^2 \text{ mm}^2 = 100 \text{ mm}^2$ $1 \text{ m}^2 = 100^2 \text{ cm}^2 = 10\,000 \text{ cm}^2$

  Volume: $1 \text{ cm}^3 = 10^3 \text{ mm}^3 = 1000 \text{ mm}^3$ $1 \text{ m}^3 = 100^3 \text{ cm}^3 = 1\,000\,000 \text{ cm}^3$

Chief Examiner Says

If you are asked to change the units of area or volume, it is often easiest to change the lengths you are given to the new units and to work out the area or volume using those. If you are not given the lengths, you need to use the information in the list above.

Test Yourself (1)

Change these units.

a) $10\,500 \text{ cm}^2$ to m^2

b) 4.2 cm^3 to mm^3

More Practice S17

Bounds of measurement

- A length measured as 45 cm to the nearest centimetre lies between 44.5 cm and 45.5 cm. So the lower bound is 44.5 cm and the upper bound is 45.5 cm.

Chief Examiner Says

A common error is to give the upper bound as 45.49 cm in an example like this, since the upper bound cannot be reached. The correct bound, however, is 45.5 cm here.

Test Yourself (2)

Give the bounds of these measurements.

a) 86 mm to the nearest millimetre

b) 48 g to the nearest gram

c) 52 litres to the nearest litre

d) 80 cm to the nearest centimetre

More Practice S18

Solutions

Test Yourself (1)

a) $10\,000 \text{ cm}^2 = 1 \text{ m}^2$
   so $10\,500 \text{ cm}^2 = 10\,500 \div 10\,000$
         $= 1.05 \text{ m}^2$

b) $1 \text{ cm}^3 = 1000 \text{ mm}^3$
   so $4.2 \text{ cm}^3 = 4.2 \times 1000$
         $= 4200 \text{ mm}^3$

Test Yourself (2)

a) 85.5 mm and 86.5 mm

b) 47.5 g and 48.5 g

c) 51.5 litres and 52.5 litres

d) 79.5 cm and 80.5 cm

Compound measures

The following are examples of compound measures.

- Speed $\left(= \dfrac{\text{Distance}}{\text{Time}}\right)$
- Density $\left(= \dfrac{\text{Mass}}{\text{Volume}}\right)$
- Population density $\left(= \dfrac{\text{Population}}{\text{Area}}\right)$

Test Yourself (1)

Jane drives for 2 hours at 70 mph and then for 30 minutes at 40 mph.

a) Find the total distance for the journey.

b) Calculate her average speed for the journey.

Chief Examiner Says

The units tell you which way to divide. Density measured in g/cm^3 means grams (mass) divided by cm^3 (volume).

Chief Examiner Says

Don't be tempted to find the average of the two speeds – the times are different.

More Practice S19

Here is an exam question ...

Vivek cycled at 12 miles per hour for 12 minutes. How far did he go?

[2]

... and its solution

12 minutes $= \dfrac{12}{60} = 0.2$ hours

Distance $=$ Speed $\times$ Time
$= 12 \times 0.2$
$= 2.4$ miles

Now Try These Exam Questions

1 P is 8 km from O on a bearing of 037° and Q is 7 km due east of O.

a) Make a scale drawing showing O, P and Q. Use a scale of 1 cm to 2 km.

b) Find the distance between P and Q.

c) Find the bearing of P from Q. [5]

2 A rectangle has dimensions 354 cm by 64 cm.

a) Work out the area

   i) in cm^2.   **ii)** in m^2.

b) The dimensions were measured to the nearest centimetre.
Write down the bounds between which the dimensions must lie. [5]

3 A block of wood is a cuboid measuring 6.5 cm by 8.2 cm by 12.0 cm.

a) Calculate the volume of the cuboid.

The density of the wood is $1.5\,g/cm^3$.

b) Calculate the mass of the block. [4]

More Exam Practice SE4

Solutions

Test Yourself (1)

a) Distance $=$ Speed $\times$ Time
At 70 mph she travels $70 \times 2 = 140$ miles.
At 40 mph she travels $40 \times 0.5 = 20$ miles
Total distance $= 140 + 20$
$= 160$ miles

30 minutes $= 0.5$ hours

b) Total time $= 2.5$ hours

Average speed $= \dfrac{\text{Total distance}}{\text{Total time}}$
$= \dfrac{160}{2.5}$
$= 64$ mph

Constructions

Constructing a triangle

- You need to be able to construct a triangle given one of four different sets of information.

  - Three sides
    - **Step 1:** Draw line AB of given length.
    - **Step 2:** Use compasses to construct arcs AC and BC with the compasses set to the given lengths.
    - **Step 3:** Draw AC and BC.

  - Two sides and the angle between them
    - **Step 1:** Draw line AB of given length.
    - **Step 2:** Measure the angle at A.
    - **Step 3:** Draw line AC of given length.
    - **Step 4:** Join C to B.

  - Two angles and one side
    - **Step 1:** Draw line AB of given length.
    - **Step 2:** Measure angles at A and B.
    - **Step 3:** Draw lines AC and BC.

  - Two sides and an angle
    - **Step 1:** Draw line AB of given length.
    - **Step 2:** Measure angle at A.
    - **Step 3:** Draw a line from A towards C.
    - **Step 4:** To fix C, draw an arc at B, the radius of the arc being the second given length. This construction may give two possible triangles, ABC or ABC'.

- You can use similar methods to construct quadrilaterals.

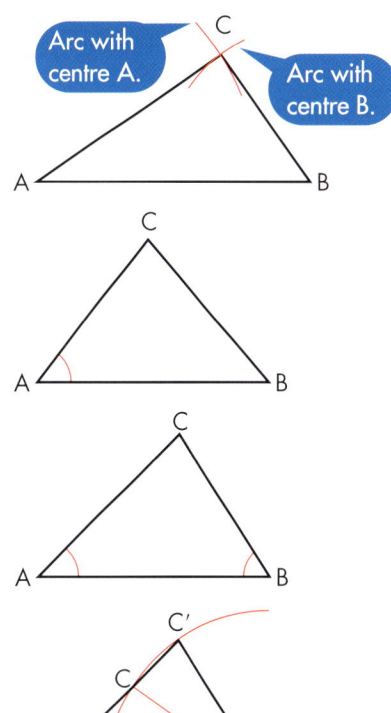

Arc with centre A.

Arc with centre B.

Test Yourself (1)

Construct an isosceles triangle ABC with sides AB = 4 cm, BC = 5 cm and AC = 5 cm. Measure angle ABC of the triangle.

Solutions

Test Yourself (1)

- **Step 1:** Draw the 4 cm side of the triangle and label it AB.
- **Step 2:** Set your compasses to 5 cm. With the compass point at A, make arcs above the centre of the line. Repeat with the point at B so that the arcs cross. This is point C.
- **Step 3:** Draw AC and BC to complete the triangle.

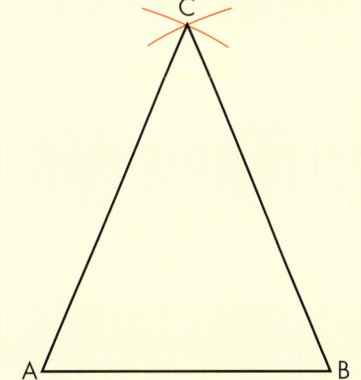

An angle of 65° to 68° is acceptable.   Angle ABC = 66°

Test Yourself (1)

Construct the quadrilateral ABCD where AB = 4.5 cm, angle BAD = 98°, angle ABC = 72°, BC = 3.4 cm and CD = 5.1 cm. Measure side AD of the quadrilateral.

More Practice S20

Other standard ruler and compasses constructions

You need to be able to do these constructions, using ruler and compasses.

● Perpendicular bisector of a line

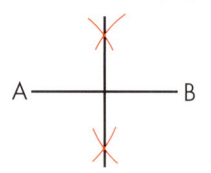

AB is the given line.

Step 1: Draw two arcs of the same radius above and below the line, centred on A and B.

Step 2: Join the intersections of the arcs. This line is the perpendicular bisector of the line AB.

● Perpendicular from a point to a line

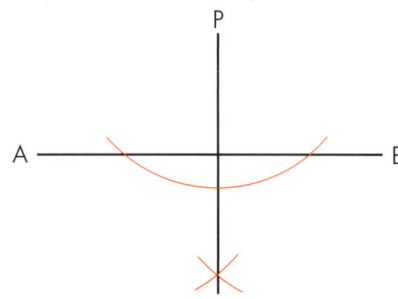

AB is the given line. P is the given point.

Step 1: Draw an arc, centred on P, to cut AB twice.

Step 2: Draw two more arcs of the same radius, centred on the points of intersection of the first arc and AB, to cut each other below the line (on the opposite side from P).

Step 3: Join the point of intersection of the two arcs to P. This line is perpendicular to AB.

● Bisector of an angle

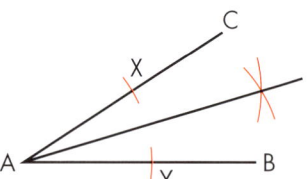

AB and AC are the two lines making the given angle.

Step 1: Draw two arcs of the same radius, centred on A, to cut AB and AC.

Step 2: Draw two more arcs of the same radius, centred on X and Y, to cut each other between AB and AC.

Step 3: Join the point of intersection of the two arcs to A. This line is the bisector of angle A.

Test Yourself (2)

Draw any triangle. Use ruler and compasses to bisect all its angles.

Chief Examiner Says

If your construction is accurate, you should find that all the bisectors meet at one point. (This point is, in fact, the centre of a circle which touches each of the three sides of the triangle tangentially – you can check by drawing this circle, if you wish.)

More Practice S21

Solutions

Test Yourself (1)

Step 1: Draw the 4.5 cm side of the quadrilateral and label it AB.

Step 2: Measure angles at A and B so that angle BAD = 98°, angle ABC = 72°.

Step 3: Measure 3.4 cm from B and mark C on the side you have drawn.

Step 4: Set your compasses to 5.1 cm and with centre C draw an arc to cut the side you drew at A (you may need to extend the line). This is point D.

Step 5: Join sides AD and CD.

Test Yourself (2)

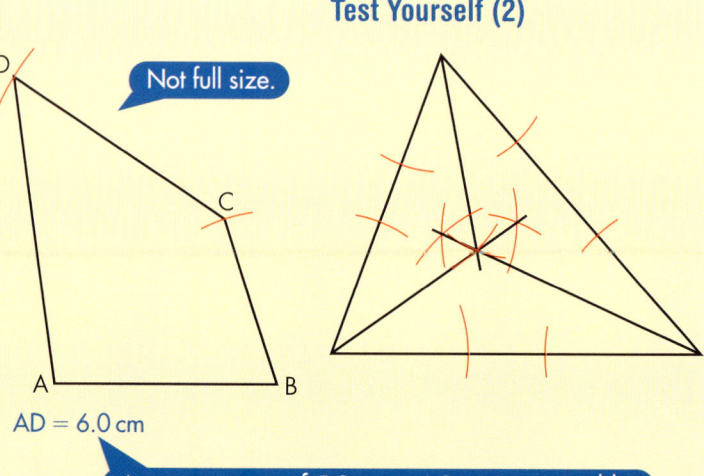

Not full size.

AD = 6.0 cm

A measurement of 5.9 cm to 6.3 cm is acceptable.

Loci

A locus is a line or region where a point can be according to a rule.

- The locus of a point which is always 3 cm from a fixed point A is a circle, centre A, radius 3 cm.

- The locus of a point which is always less than 3 cm from a fixed point A is the region inside a circle, centre A, radius 3 cm.

- The locus of a point which is always more than 3 cm from a fixed point A is the region outside a circle, centre A, radius 3 cm.

- The locus of a point which is 2 cm from a fixed straight line (or line segment) is one of a pair of lines parallel to that line.

- The locus of a point which is an equal distance from two fixed points is the perpendicular bisector of the line joining the two points.

- The locus of a point which is an equal distance from two fixed lines is one or both bisectors of the angles between the lines.

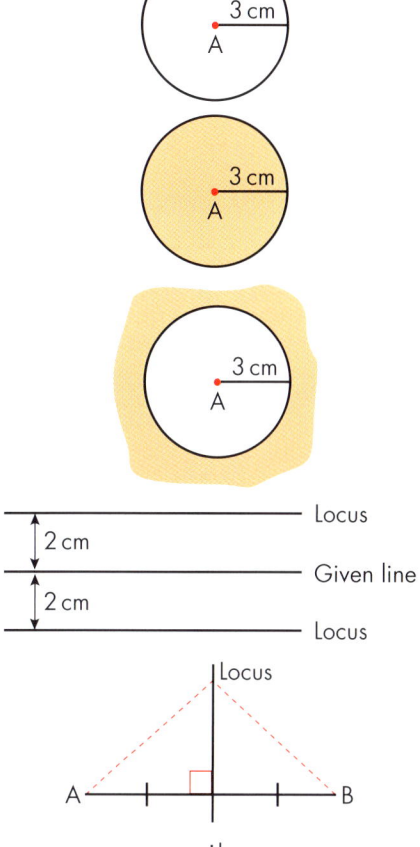

Test Yourself (1)

A goat is tethered by a 2 m rope to a rail 3 m long, which is fixed in a field of grass. The rope can slide along the rail. Draw a scale diagram to show the region of grass which the goat can eat.

More Practice S22

Chief Examiner Says

Often, exam questions involve more than one locus. Solve these questions in stages, drawing one locus at a time. Show the final required region or points clearly by labelling, together with shading if necessary.

Solutions

Test Yourself (1)

Scale: 1 cm to 1 m

Here is an exam question ...

a) Two buoys are anchored at A and B. B is due East of A.
A boat is anchored at C.

  i) Using a scale of 1 cm to 2.5 m, draw the triangle ABC.   **[2]**

  ii) Measure the bearing of the boat, C, from buoy A.   **[2]**

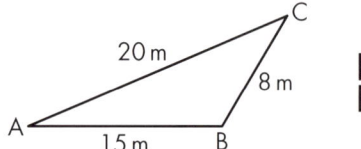

b) This is the plan of a garden drawn on a
scale of 1 cm to 2 m.
A pond is to be dug in the garden.

The pond must be at least 4 m from the tree.
It must be at least 3 m from the house.
Shade the region where the pond can be
dug. Show all your construction lines.

   × Tree

   House

   [3]

... and its solution

a) i) Step 1: Draw the line AB 6 cm long.

     Step 2: Using compasses, draw an arc 8 cm from A,
and an arc 3.2 cm from B.

     Step 3: Mark the point C where the arcs cross and
join to A and B to complete the triangle.

ii) To measure the bearing, use your
protractor, to draw the North line
at A, at right-angles to AB.
Now use your protractor, with the
zero line along the North line, to measure the bearing. It should be between 069° and 070°.

b) At least 4 m from the tree means it is outside
a circle radius 2 cm, centre the tree.
At least 3 m from the house means it is to
the left of a line parallel to the house and
1.5 cm from it.

Scale 1 cm to 2 m

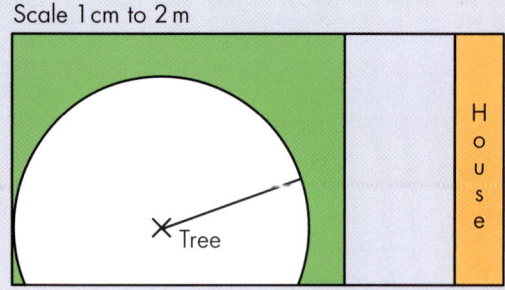

Now Try These Exam Questions

1 The diagram shows a triangle ABC.
The bisector of the angle at A meets line
BC at X.

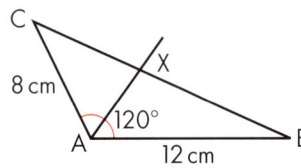

  a) Construct the triangle and the bisector of
angle A.

  b) Measure the distance AX.   **[5]**

2 Ashwell and Buxbourne are two towns, 50 km
apart. Chris is house-hunting. He has decided
he would like to live closer to Buxbourne than
Ashwell but no further than 30 km from
Ashwell.
Using a scale of 1 cm to represent 5 km,
construct and shade the area in which Chris
should look for a house.   **[4]**

More Exam Practice SE5

Identification and selection of data

Primary and secondary data

- Primary data is data which is collected by the person using the information.
- Secondary data is data obtained from elsewhere such as books or databases.

Grouped data and class intervals

- When handling lots of data it is useful to collect the data in groups.
- Often grouped data is put into 'equal class intervals' – that is into groups of equal width.
- As a rough guide about 5 to 10 groups is usually about right. If in doubt use smaller intervals to start with as you can always combine intervals later.
- Make sure your intervals do not overlap.

Solutions

Test Yourself (1)

a) Secondary
b) Primary

Test Yourself (2)

a)

Time (min)	Tally	Frequency
21–25	||||	4
26–30	||||	5
31–35	||||	4
36–40	|||| |	6
41–45	||	2
46–50	||	2
51–55	||||	5
56–60	||	2

b)

Time (min)	Frequency
21–30	9
31–40	10
41–50	4
51–60	7

Surveys

- Questions you ask should not be biased nor should they be leading questions.
- Questions should give a choice of answers. This avoids a large number of vague answers that are difficult to analyse.
- Questions should not cause embarrassment.
- Questions should be short, clear, easy to understand and relevant.
- You should ask a variety of people at a variety of times in a variety of places so that your sample of people is unbiased – that is representative of the population as a whole.

Test Yourself (1)

Criticise this question in a survey.

'Normal people enjoy watching EastEnders.

Do you watch EastEnders?'

More Practice HD3

Here is an exam question ...

Here are two questions that Amy wanted to include in a survey she was going to do to find out if people liked the new shopping centre in her town.

Rewrite each one to show how you would word the questions in a questionnaire and why you would change it.

a) How old are you? **[2]**

b) This new shopping centre appears to be a success. Do you agree? **[2]**

... and its solution

Possible answers would be:

a) The question may be thought to be personal – some people may not answer.
Change to:
What is your age? Tick the appropriate box.

☐ 10–19 ☐ 20–29 ☐ 30–39

☐ 40–49 ☐ 50–59 ☐ 60 or over

b) This is a leading question. Change to:
Do you think the shopping centre is a success? Tick the appropriate box.

☐ yes ☐ no ☐ don't know

Now Try These Exam Questions

1 Wasim, Abbie and Oliver carried out a survey to find the number of cars passing their school in ten-minute intervals over a period of six hours. The results are shown below.

```
14  21  30  26  51  39  31  15   8
17  20  16  18  25  16  34  28  26
 7  12   5  44  36  58  25  22  24
17  42  13  44  28  33  16  27  26
```

To analyse their results they each decided to group their data and make a frequency table.

Wasim chose these groups:
0–19, 20–39, 40–59.

Abbie chose these groups:
0–10, 10–20, 20–30, 30–40, 40–50, 50–60.

Oliver chose these groups:
0–9, 10–19, 20–29, 30–39, 40–49, 50–59.

Their teacher said that Oliver's was the best method.

a) Give one reason each for why she thought Wasim's and Abbie's groups were unsuitable. **[2]**

b) Copy and complete the following frequency table using Oliver's groups of numbers of cars. You may want to use tally marks.

Cars	Frequency
0–9	
10–19	
20–29	
30–39	
40–49	
50–59	

[2]

More Exam Practice HDE1

Solutions

Test Yourself (1)

This is a biased question because if you do not watch 'soaps' in general or EastEnders in particular, it does not mean that you are not normal.

Processing and representing data

Frequency diagrams and polygons

- For grouped information with equal intervals, a frequency diagram can be drawn with the heights of the bars equal to the frequency.
- A frequency polygon is formed by joining, with straight lines, the midpoints of the tops of the bars in a frequency diagram.

Test Yourself (1)

The table shows the amount of pocket money received by the members of class 10h.

Amount (£)	0–3.99	4–7.99	8–11.99	12–15.99
Frequency	3	10	12	5

To show this information, draw the following.

a) A frequency diagram   **b)** A frequency polygon.

More Practice HD4

Scatter diagrams and lines of best fit

- Scatter diagrams (or graphs) are drawn to investigate any possible link between two variables. Values of the two variables are plotted as points on a graph. If the points lie approximately in a straight line than there is a **correlation** between the two variables. The closer the points are to a straight line, the stronger the correlation.

- The correlation is **positive** if the higher the value of x, the higher the value of y.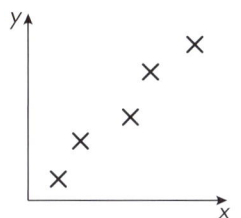

- The correlation is **negative** if the higher the value of x, the lower the value of y.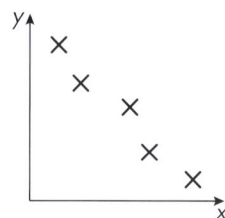

- If there is a correlation, then a **line of best fit** can be drawn. It should reflect the slope of the points and have approximately the same number of points on either side. It can be used to estimate a value of y from a given value of x. It should not be used to estimate too far outside the range of the data.

Test Yourself (2)

The table shows the time spent and marks gained for 10 students' coursework.

Time (h)	5	8	3	6	6	7	4	10	8	7
Mark	12	15	9	14	11	13	10	19	16	14

a) Plot a scatter diagram for this information.   **b)** What conclusions can be drawn?

c) Draw a line of best fit.

d) Use your line of best fit to estimate the mark of a student who spent nine hours on the coursework.

More Practice HD5

Solutions

Test Yourself (1)

a) **b)**

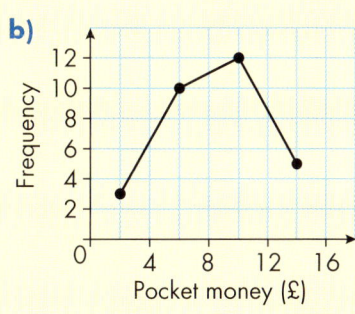

Test Yourself (2)

a) c)

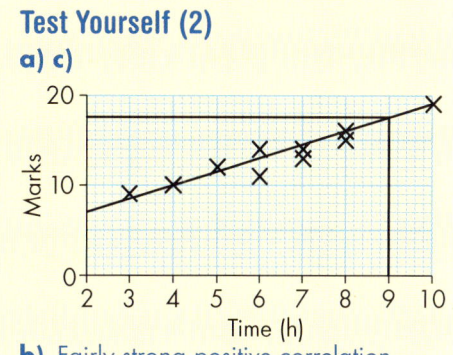

b) Fairly strong positive correlation
d) 17 or 18

Mean from a frequency distribution

- To calculate the mean from a frequency distribution:
  - **i)** Multiply each observation (x) by its corresponding frequency
  - **ii)** Then add up these results
  - **iii)** Finally divide by the total frequency
- For grouped data, use the middle of the interval for the value of x to represent each group.

> **Chief Examiner Says**
>
> You are often told the total frequency in the question.

Test Yourself (1)

a) The table shows the number of letters delivered one morning to a street of 100 houses. Find the mean number of letters.

No. of letters (x)	0	1	2	3	4	5
Frequency	8	19	28	25	17	3

b) Find the mean amount of pocket money received by the students in class 10h on page 49.

More Practice HD6

Stem-and-leaf diagrams and median

- Large numbers of two-digit data items can be displayed in a stem-and-leaf diagram. The tens digit (the stem) is written in the left-hand column and the units digits (the leaves) are written in the right-hand column.
- It is best to draw a stem-and-leaf diagram, initially with 'leaves' unordered, and then redraw it with the 'leaves' ordered.
- You can also use a stem-and-leaf diagram, for example, with the units as the 'stem' and the first decimal place as the 'leaves'.
- You should always give a key, for example 3 | 2 means 32
- It is fairly easy to then find the median by counting up to the middle value.

Test Yourself (2)

Here are the test marks for a group of students.

63	54	34	42
38	76	69	52
45	54	67	71
64	43	57	59
47	66	57	63

a) Draw a stem-and-leaf diagram for these results. Don't forget the key.

b) Use your stem-and-leaf diagram to find the median score.

More Practice HD7

Solutions

Test Yourself (1)

a) Mean = $\dfrac{0 \times 8 + 1 \times 19 + 2 \times 28 + 3 \times 25 + 4 \times 17 + 5 \times 3}{100}$ = 233 ÷ 100 = 2.33

b)

Mid-interval value (x)	2	6	10	14
Frequency	3	10	12	5

Mean = $\dfrac{2 \times 3 + 6 \times 10 + 10 \times 12 + 14 \times 5}{3 + 10 + 12 + 5}$ = 256 ÷ 30 = £8.53

> **Chief Examiner Says**
>
> Check that the total frequency matches that given in the question.

Test Yourself (2)

a) Unordered

```
3 | 4  8
4 | 2  5  3  7
5 | 4  2  4  7  9  7
6 | 3  9  7  4  6  3
7 | 6  1
```

Ordered

```
3 | 4  8
4 | 2  3  5  7
5 | 2  4  4  7  7  9
6 | 3  3  4  6  7  9
7 | 1  6
```

Key: 3 | 4 means 34

b) Since there are 20 marks in the group there are two middle values, the 10th and 11th. They are both 57, so the median is 57.

Here is an exam question ...

A wedding was attended by 120 guests. The distance, d miles, that each guest travelled was recorded in the frequency table.

Distance (d miles)	Number of guests
$0 < d \leq 10$	26
$10 < d \leq 20$	38
$20 < d \leq 30$	20
$30 < d \leq 50$	20
$50 < d \leq 100$	12
$100 < d \leq 140$	4

Calculate an estimate of the mean distance travelled. **[4]**

... and its solution

> Use the mid-interval values, 5, 15,... to represent each group.

$$\text{Mean} = \frac{5 \times 26 + 15 \times 38 + ... + 120 \times 4}{120}$$

$$= 3380 \div 120$$

$$= 28.2 \text{ miles}$$

Now Try These Exam Questions

1 A class of 33 students sat a mathematics exam. Their results are listed below.

```
89  78  56  43  92  95  24  72  58
65  55  98  81  72  61  44  48  76
82  91  76  81  74  82  99  21  34
79  64  78  81  73  69
```

a) Draw an ordered stem-and-leaf diagram for this information. **[3]**

b) Find the median mark. **[1]**

2 The number of matches in each of 100 boxes was counted. Here are the results.

Number of matches	45	46	47	48	49	50
Frequency	14	25	32	19	8	2

Calculate the mean number of matches in a box. **[3]**

More Exam Practice HDE2

Comparing distributions

- If you are asked to compare two distributions, make two basic comparisons.
- For the first comparison, decide which distribution is bigger on average. Evidence of this is a higher mean or a higher median or a higher mode. In a frequency diagram or polygon, a higher mode is indicated by the highest bar or highest point being further to the right.
- For the second comparison, decide which distribution has the greater spread. This is indicated by a bigger range. Later you will meet another measurement of spread called the inter-quartile range.

Test Yourself (1)

The stem-and-leaf diagrams below show the marks scored by classes 10A and 10B in their last Maths test.

```
10A                          10B

2 | 5 6 7                     1 | 1 8
3 | 2 2 3 4 6 8              2 | 2 5
4 | 0 1 3 6 6 7 8 9         3 | 3 6 7
5 | 3 4 6 8 9               4 | 2 5 5 6 7 8
6 | 4 4 5 6                 5 | 4 6 6 7 8 9 9
7 | 0 1 1                   6 | 0 2 3 5 6 8
                            7 | 1 3 4
```

Key: 3|2 means 32

Make two comparisons between the two distributions.

More Practice HD8

Solutions

Test Yourself (1)

10A: Median (15th mark) = 47   Range = 71 − 25 = 46

10B: Median (15th mark) = 56   Range = 74 − 11 = 63

So class 10B have the higher scores on average (median) and class 10B have greater spread of scores (range).

Here is an exam question ...

The manager of the Metro cinema records the numbers of people watching each of two films for 25 days. The frequency diagram is for Film A. The table shows the numbers of people who watched Film B. Compare the two distributions.

[2]

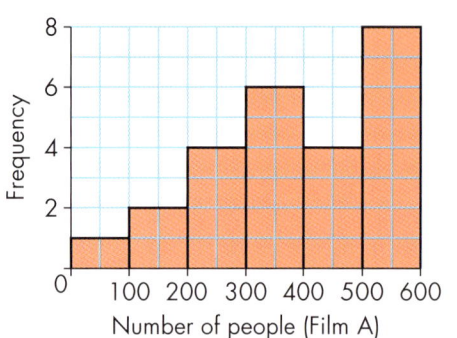

Number of people (Film B)	Frequency
0–99	5
100–199	12
200–299	6
300–399	1
400–499	0
500–599	1

... and its solution

The average attendance at Film A was much higher, or more people watched Film A. The numbers attending Film A were more varied or the number watching Film B each night was more consistent.

Now Try This Exam Question

1   A botanist collected samples of leaves from two oak trees.
    The stem-and-leaf diagrams show the lengths in centimetres.

Tree A

```
5 | 5 7
6 | 0 2 2 3 5 7 8
7 | 1 3 8 9
8 | 2 2 3 8
9 | 4 5 7
```

Tree B

```
5 |
6 | 4 5 6 6 7 9
7 | 0 0 1 2 4 5 5 6 7 8
8 | 1 1 3 7
9 |
```

Key: 6|4 = 6.4 cm

Compare the two distributions.   [2]

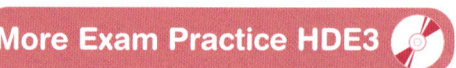

More Exam Practice HDE3

Probability

Basic probability

- Probability can be expressed as fractions, decimals or percentages. All probabilities are between 0 and 1 inclusive.
- P(A does not happen) = 1 − P(A).

  P(A) is a useful shorthand for the probability that A happens.

- For equally likely outcomes

  $$P(A) = \frac{\text{number of ways A can happen}}{\text{total possible number of outcomes}}$$

- Mutually exclusive outcomes are those which cannot happen together. If A, B and C are mutually exclusive outcomes covering all the possibilities, then P(A) + P(B) + P(C) = 1

Test Yourself (1)

A jar contains red, green and white beads. One bead is selected at random. The probability that the bead is red is 0.3. The probability that the bead is green is 0.5. What is the probability that the bead is white?

More Practice HD9

Relative frequency

- Relative frequency = $\dfrac{\text{number of times event occurs}}{\text{total number of trials}}$

- When theoretical probabilities are not known, relative frequency can be used to estimate probability.
- The greater the number of trials, the better the estimate. The graph of relative frequency against the number of trials may vary greatly at first, but later 'settles down'.
- Relative frequency experiments may be used to test for bias, for example to see if a dice is fair.

Test Yourself (2)

Sarah has a biased dice. This graph shows the relative frequency of throwing a 6 when she threw the dice 1000 times.

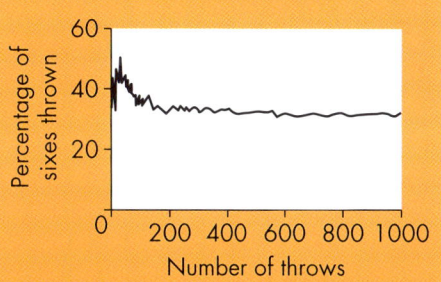

Use the graph to estimate the probability of throwing a 6 with Sarah's dice. Explain your answer.

More Practice HD10

Solutions

Test Yourself (1)

P(R) + P(G) = 0.3 + 0.5 = 0.8
P(W) = 1 − 0.8 = 0.2

Chief Examiner Says

Never write probabilities as '1 in 5' or as a ratio '1 to 5' or 1:5. You will lose marks if you do. Instead, write $\frac{1}{5}$ or 0.2 (or 20% if the question uses percentages).

Test Yourself (2)

Since Sarah has thrown the dice a large number of times she can use the result from 1000 throws as a probability.

From the graph this is about 30% or 0.3.

Here is an exam question ...

The cards used in a child's game have either a square, a triangle, a circle or a star printed on them.

The table shows the probabilities of choosing the shapes.

Outcome	Square	Triangle	Circle	Star
Probability	0.2	0.35		0.3

a) Copy and complete the table. **[2]**
The cards are either red or blue. There are three times as many blue cards as red cards.

b) What is the probability that the card drawn is red? **[2]**

... and its solution

a) All the probabilities must add to 1.
0.2 + 0.35 + 0.3 = 0.85
Therefore the probability for the circle cards is 0.15.

b) 3 'parts' of the total are blue so 1 'part' is red. The total number of 'parts' is 4.
So the probability that the card is red is $\frac{1}{4}$.

Now Try These Exam Questions

1 Ahmed is counting vehicles passing a junction between 8.00 a.m. and 8.30 a.m.

Vehicle	Frequency
Cars	72
Motorcycles	15
Lorries	28
Vans	33
Buses	12

a) Use these data to find the probability that the next vehicle to pass the junction.
  i) is a car. **[2]**
  ii) is a bus. **[1]**
  iii) has more than two wheels. **[2]**

Give your answers as fractions in their lowest terms.

b) Will this give reliable results for vehicles passing the junction at 11.00 p.m.? Explain your answer. **[1]**

2 There are only red, green and blue counters in a bag.

a) Copy and complete the table to show the probability of choosing at random a blue counter from the bag.

Colour	Red	Green	Blue
Probability	0.2	0.45	

b) A counter is chosen from the bag, its colour is noted and then replaced.
If this is repeated 200 times, how many counters would you expect to be green? **[2]**

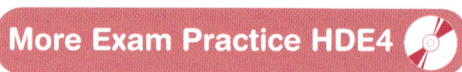
More Exam Practice HDE4

Powers and roots

Indices which are integers

- Indices (or powers) give a short way of writing numbers multiplied by themselves.

  For example

  $$2^3 = 2 \times 2 \times 2 = 8 \qquad 2^0 = 1 \qquad 2^{-1} = \frac{1}{2^1} = \frac{1}{2}$$

Chief Examiner Says

Don't forget that $n^0 = 1$ and $n^{-a} = \dfrac{1}{n^a}$ for any non-zero value of n.

Test Yourself (1)

Find the value of the following.

a) 4^2 b) 5^3

c) 3^{-2} d) 7^0

e) 2^{-4}

More Practice N25

Exponential growth and decay

- Use a constant multiplier – greater than 1 for growth, less than 1 for decay.
- Use the x^y or $\land$ key.

Test Yourself (2)

a) A car depreciates by 30% every year. It cost £9000 new. How much is it worth after 5 years?

b) A population of bacteria increases by 6% every hour. By what factor has the population grown after 24 hours?

More Practice N26

Multiplying and dividing with indices

- Multiplying: $n^a \times n^b = n^{a+b}$ Add the powers
- Dividing: $\dfrac{n^a}{n^b} = n^{a-b}$ Subtract the powers
- Powers of powers: $(n^a)^b = n^{ab}$ Multiply the powers

More Practice N27

Test Yourself (3)

Simplify the following.

a) $3^2 \times 3^4$ b) $2y^4 \times 3y^5$

c) $\dfrac{3^5}{3^3}$ d) $(2^3)^4$

Solutions

Test Yourself (1)

a) 16
b) 125
c) $\frac{1}{9}$
d) 1
e) $\frac{1}{16}$

Test Yourself (2)

a) $9000 \times (0.7)^5$
  $= £1512.63$

b) $(1.06)^{24} = 4.0489...$
  Approximately 4 times.

Test Yourself (3)

a) 3^6 $2 + 4 = 6$

b) $6y^9$ $4 + 5 = 9$

c) 3^2 $5 - 3 = 2$

d) 2^{12} $3 \times 4 = 12$

Chief Examiner Says

For adding and subtracting, indices are not so helpful. There is no shortcut rule.

So, for example
$3^3 + 3^2 = 27 + 9 = 36$.

Indices which are fractions

- $n^{1/2} = \sqrt{n}$
- $n^{1/a} = \sqrt[a]{n}$
- $n^{a/b} = (n^{1/b})^a$ or $(\sqrt[b]{n})^a$

More Practice N28

Test Yourself (1)

Simplify.

a) $64^{1/3}$ b) $125^{-2/3}$

Standard index form

- This is used for very large and very small numbers.
- They are written in the form $a \times 10^n$, where n is an integer and $1 \leqslant a < 10$
- To multiply or divide numbers in standard form without a calculator, use the rules of indices. For example
  $(3 \times 10^6) \times (5 \times 10^8) = 15 \times 10^{6+8} = 15 \times 10^{14} = 1.5 \times 10^{15}$

More Practice N29

Test Yourself (2)

1 Write these numbers in standard form.

a) 3 500 000

b) 0.000 42

c) 43.2

d) 0.03

2 Write these as ordinary numbers.

a) 4×10^3

b) 2.74×10^4

c) 5.3×10^{-1}

d) 4.68×10^{-4}

Here is an exam question ...   ... and its solution

a) Simplify.

  i) $\dfrac{(pq^3)^2}{p^3q}$ **ii)** $(p^8)^{-3/4}$ [5]

b) Write these numbers in standard form.
  i) 41 000 000   **ii)** 0.000 062 9   [2]

c) Work these out. Give your answer in standard form.
  i) $(3 \times 10^{-9}) \times (7 \times 10^{11})$
  ii) $\dfrac{6 \times 10^4}{2 \times 10^{-4}}$ [4]

a) i) $\dfrac{p^2q^6}{p^3q} = p^{-1}q^5 = \dfrac{q^5}{p}$

For p: $2 \times 1 - 3 = -1$
For q: $3 \times 2 - 1 = 5$

  ii) $p^{-6} = \dfrac{1}{p^6}$

$8 \times -\dfrac{3}{4} = -6$

b) i) 4.1×10^7
  ii) 6.29×10^{-5}

$7 \times 3 = 21; -9 + 11 = 2$

c) i) $21 \times 10^2 = 2.1 \times 10^3$
  ii) 3×10^8

$\dfrac{6}{2} = 3; 4 - -4 = 8$

Solutions

Test Yourself (1)

a) $\sqrt[3]{64} = 4$

b) $(\sqrt[3]{125})^{-2} = 5^{-2} = \dfrac{1}{25}$

Chief Examiner Says

Know how to obtain the answers using the root and power buttons on your calculator, as well as without a calculator.

Test Yourself (2)

1 a) 3.5×10^6

  b) 4.2×10^{-4}

  c) 4.32×10^1

  d) 3×10^{-2}

2 a) 4000

  b) 27 400

  c) 0.53

  d) 0.000 468

Chief Examiner Says

To enter standard form numbers on a calculator, use the EXP or EE button.

e.g. for 3.5×10^6 enter

Now Try These Exam Questions

1 Write the following as whole numbers or fractions.

a) 9^{-2}

b) 9^0

c) $27^{1/3}$ [3]

2 a) Work out $(3.0 \times 10^4) \times (6.0 \times 10^3)$, writing the answer in standard form.

b) A terawatt is 10^{12} watts. A power station produces 1.2×10^8 watts. Write this in terawatts.   [3]

More Exam Practice NE8

Fractions and decimals

Multiplying and dividing fractions

- When mixed numbers are involved, first change to improper fractions.
- To multiply fractions, multiply the numerators and denominators and then simplify.
- To divide fractions, invert the second fraction and multiply.

Test Yourself (1)

Work out these.

a) $2\frac{1}{2} \times 1\frac{2}{5}$

b) $3\frac{3}{4} \div \frac{1}{2}$

More Practice N30

Increasing or decreasing by a fraction

To increase or decrease an amount by a fraction:
- Add or subtract the fraction to or from 1.
- Multiply by the new fraction.

Test Yourself (2)

Increase 18 by $\frac{1}{3}$.

More Practice N31

Solutions

Test Yourself (1)

a) $2\frac{1}{2} \times 1\frac{2}{5} = \frac{5}{2} \times \frac{7}{5}$ Make fractions improper.

$= \frac{1}{2} \times \frac{7}{1}$ Cancel by 5

$= \frac{7}{2} = 3\frac{1}{2}$

Make fraction improper.

b) $3\frac{3}{4} \div \frac{1}{2} = \frac{15}{4} \div \frac{1}{2}$

$= \frac{15}{{}_2 4} \times \frac{2^1}{1}$ Invert the second fraction and multiply.

$= \frac{15}{2} \times \frac{1}{1}$ Cancel by 2

$= \frac{15}{2} = 7\frac{1}{2}$

Test Yourself (2)

$18 \times \frac{4}{3} = 24$

$1 + \frac{1}{3} = \frac{4}{3}$

Finding an amount before an increase or decrease

To find the amount before an increase or decrease:
- Add or subtract the fraction to or from 1.
- Divide by the fraction.

More Practice N32

Test Yourself (1)

After an increase of $\frac{1}{5}$, the cost of a coat is £54. What did it cost before the increase?

Recurring and terminating decimals

- All fractions are equal to either recurring or terminating decimals.
- The fractions equal to terminating decimals have denominators whose only prime factors are 2 and/or 5.

Test Yourself (2)

Write these fractions as decimals.

a) $\frac{1}{8}$ **b)** $\frac{1}{6}$

c) $\frac{13}{20}$ **d)** $\frac{7}{25}$

e) $\frac{3}{7}$

Changing a recurring decimal to a fraction

- The method for doing this is best illustrated by an example.
  Express $0.\dot{4}\dot{2}$ as a fraction in its lowest terms.
  Let $r = 0.\dot{4}\dot{2}$
  So
  $$r = 0.424\,242\,42\ldots$$
  $$100r = 42.424\,242\,42\ldots$$
  Subtracting $99r = 42$
  $$r = \frac{42}{99} = \frac{14}{33}$$

Test Yourself (3)

Express $0.\dot{5}0\dot{7}$ as a fraction in its lowest terms.

More Practice N33

Chief Examiner Says

Multiply by 10^n, where n is the number of recurring figures.

Solutions

Test Yourself (1)
Old price = £54 ÷ $1\frac{1}{5}$ = £54 ÷ $\frac{6}{5}$ = £54 × $\frac{5}{6}$ = £45

Test Yourself (2)
a) 0.125 — Terminating, prime factor 2

b) $0.1\dot{6}$ — Recurring, prime factors 2 and 3

c) 0.65 — Terminating, prime factors 2 and 5

d) 0.28 — Terminating, prime factor 5

e) $0.\dot{4}2857\dot{1}$ — Recurring, prime factor 7

Test Yourself (3)
Let $r = 0.\dot{5}0\dot{7}$
So
$$r = 0.507\,507\,507\ldots$$
$$1000r = 507.507\,507\,507\ldots$$
Subtracting $999r = 507$
$$r = \frac{507}{999} = \frac{169}{333}$$

Here is an exam question ...   ... and its solution

The Midland Railway has a special offer on some fares. This is the advertisement:

Fares Reduced
$\frac{1}{3}$ off normal fare

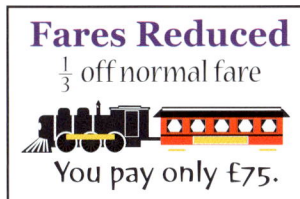

You pay only £75.

What is the cost of the normal fare?   **[3]**

Fare is reduced by $\frac{1}{3}$.

$$\text{Normal fare} = \frac{75}{1 - \frac{1}{3}} = \frac{75}{\frac{2}{3}}$$

$$= 75 \times \frac{3}{2} = £112.50$$

Now Try These Exam Questions

1 Work out the following. Give your answers as fractions as simply as possible.

 a) $1\frac{3}{5} \times 2\frac{2}{9}$ **b)** $1\frac{1}{4} + 2\frac{3}{5}$ **[6]**

2 $h = \dfrac{2xy}{x + y}$. Find h when $x = \dfrac{2}{5}$ and $y = \dfrac{2}{7}$.   **[6]**

More Exam Practice NE9

Percentages

Finding an amount before a percentage increase or decrease

To find the original amount before an increase or decrease:
- Add or subtract the percentage to or from 100
- Change to a decimal
- Divide

More Practice N34

Test Yourself (1)

The price of a skirt was reduced by 5%. It now costs £27.55. What was the original price?

Here is an exam question ...   ... and its solution

a) Jenny used to go to the village hairdresser where a trim cost £4. The hairdresser left so she went to a salon in town. A trim there cost £30. What percentage increase is this?   **[2]**

b) At the town salon the prices have gone up several times. The last increase was 6%. Jenny now pays £37.10. How much did she pay before the last increase?   **[2]**

a) Increase = £26. Original was £4.
Percentage increase $= \frac{26}{4} \times 100$
$= 650\%$

b) New = 106% of previous. (1.06)

Previous is wanted so divide by 1.06

Previous = 37.10 ÷ 1.06
$= £35$

Solutions

Test Yourself (1)

Decrease, so new price = 100 − 5 = 95% of original price.

Original price = 27.55 ÷ 0.95 = £29.

To find original price, divide by 0.95 (95% = 0.95)

Now Try These Exam Questions

1 The value of a car falls by 20% each year. A car which is 3 years old is worth £10 240.

  a) What was it worth last year? **[2]**

  b) What was it worth when it was new? **[2]**

More Exam Practice NE10

Written methods

Direct proportion

● If two quantities vary in direct proportion it means that if you multiply one quantity by a number, you multiply the other quantity by the same number.

> **Test Yourself (1)**
>
> The length of a stalactite is directly proportional to the time it has been growing. After 100 years it is 14 cm long.
> How long will it be after these number of years?
>
> **a)** 200 years
>
> **b)** 1000 years

Inverse proportion

● Inverse proportion means that if you multiply one quantity by a number, you divide the other quantity by the same number.

So $y \propto \dfrac{1}{x}$ and $y = \dfrac{k}{x}$

> **Chief Examiner Says**
>
> Direct and inverse proportion problems can also be solved using algebra. See page 67 for more work on this topic.

> **Test Yourself (2)**
>
> It takes 2 hours at 60 mph to drive between two towns. How long does it take at 50 mph?

More Practice N35

Solutions

Test Yourself (1)

a) $14 \times \dfrac{2\cancel{0}\cancel{0}}{1\cancel{0}\cancel{0}} = 14 \times 2$

$\qquad = 28$ cm

b) $14 \times \dfrac{10\cancel{0}\cancel{0}}{1\cancel{0}\cancel{0}} = 14 \times 10$

$\qquad = 140$ cm

Alternatively, you could work out the growth per year (0.14 cm) and then multiply by the number of years.

Test Yourself (2)

$2 \times \dfrac{60}{50} = 2.4$ hours

$\qquad = 2$ hours 24 minutes

Surds and π

- A surd is an expression involving a square root sign.
  For example $\sqrt{3}$ or $5 + \sqrt{2}$
- When simplifying surds, look for the highest factor that has an exact square root
- When multiplying surds, use the normal rules of algebra
- To rationalise the denominator of a fraction with a simple surd in the denominator, multiply the numerator and denominator by the surd
- If you are asked for an exact answer, leave your answer in surd form or as a multiple of π

Chief Examiner Says

On the non-calculator paper you may be told to leave answers in terms of π.

Test Yourself (1)

1 Simplify the following.
  a) $\sqrt{80}$
  b) $(2 + \sqrt{3})(5 + 2\sqrt{3})$

2 Rationalise the denominator $\dfrac{6}{\sqrt{15}}$

3 Find the exact area of a circle of radius 5.

More Practice N36

Here is an exam question ...   ... and its solution

The wavelength of radio waves, is inversely proportional to the frequency. The table shows some radio stations with their frequencies and wavelengths.

Radio station	Frequency (F)	Wavelength (L)
Radio Atlantic	252 kHz	1179 m
Virgin Radio	1215 kHz	
BBC World Service		458.5 m

a) Calculate the wavelength of Virgin Radio to the nearest metre. **[2]**

b) Calculate the frequency of the BBC World Service. **[2]**

a) $1179 \times \dfrac{252}{1215} = 245$ m

b) $252 \times \dfrac{1179}{458.5} = 648$ kHz

Now Try These Exam Questions

1 **a)** Express the following in the form $p\sqrt{q}$ where p and q are integers and q is as small as possible. For example $\sqrt{8} = 2\sqrt{2}$.
   i) $\sqrt{72}$ **ii)** $\sqrt{20} \times \sqrt{15}$
   iii) $\dfrac{\sqrt{50} \times \sqrt{27}}{\sqrt{18}}$

b) Given that $(5 + \sqrt{7})^2 = a + b\sqrt{7}$, find the values of a and b. **[8]**

2 A car uses 14 litres of fuel to travel 80 km. How much fuel will it use to travel 250 km? **[2]**

3 The value of a car is inversely proportional to its age in years. My three-year-old car is worth £9500. How much will it be worth when it is 8 years old? **[2]**

More Exam Practice NE11

Solutions

Test Yourself (1)

1 **a)** $\sqrt{80} = \sqrt{16 \times 5} = 4\sqrt{5}$

b) $(2 + \sqrt{3})(5 + 2\sqrt{3})$
$= 10 + 4\sqrt{3} + 5\sqrt{3} + 2\sqrt{3}\sqrt{3}$
$= 10 + 6 + 9\sqrt{3} = 16 + 9\sqrt{3}$

$2 \times \sqrt{3} \times \sqrt{3} = 2 \times 3 = 6$

2 $\dfrac{6 \times \sqrt{15}}{\sqrt{15} \times \sqrt{15}} = \dfrac{6\sqrt{15}}{15} = \dfrac{2\sqrt{15}}{5}$

3 Area $= \pi r^2 = \pi \times 25 = 25\pi$ cm^2

Calculator methods

Standard form

- Use the ⌷EXP⌷ or ⌷EE⌷ buttons.

  [These are in place of the $\times 10$ part of a standard form number.]

More Practice N37

Test Yourself (1)

Work out the following.

a) $2.3 \times 10^6 + 5 \times 10^7$

b) $\dfrac{6.2 \times 10^8 - 7.5 \times 10^7}{1.5 \times 10^3}$

Trigonometric keys

- These are the sin, cos and tan keys.
- You also need to be able to use the $\sin^{-1}$, $\cos^{-1}$ and $\tan^{-1}$ keys for finding angles.

Test Yourself (2)

1 Work out the following.

a) $\sin 30° + \cos 60°$

b) $\dfrac{4 + \tan 45°}{5^2}$

2 Find x if $\tan x = 1.2$.

More Practice N38

Solutions

Test Yourself (1)

 Give your answer in standard form

a) ⌷2⌷ ⌷.⌷ ⌷3⌷ ⌷EXP⌷ ⌷6⌷ ⌷+⌷ ⌷5⌷ ⌷EXP⌷ ⌷7⌷ ⌷=⌷ 5.23×10^7

b) ⌷[⌷ ⌷6⌷ ⌷.⌷ ⌷2⌷ ⌷EXP⌷ ⌷8⌷ ⌷−⌷ ⌷7⌷ ⌷.⌷ ⌷5⌷ ⌷4⌷ ⌷EXP⌷ ⌷7⌷ ⌷]⌷

⌷÷⌷ ⌷1⌷ ⌷.⌷ ⌷5⌷ ⌷EXP⌷ ⌷3⌷ ⌷=⌷ 3.64×10^5 (to 3 s.f.)

Test Yourself (2)

1 a) ⌷sin⌷ ⌷30⌷ ⌷+⌷ ⌷cos⌷ ⌷60⌷ ⌷=⌷ 1 on most calculators

or

⌷30⌷ ⌷sin⌷ ⌷+⌷ ⌷60⌷ ⌷cos⌷ ⌷=⌷ 1

b) ⌷[⌷ ⌷4⌷ ⌷+⌷ ⌷tan⌷ ⌷45⌷ ⌷]⌷ ⌷÷⌷ ⌷5⌷ ⌷x^2⌷ ⌷=⌷ 0.2

2 ⌷SHIFT⌷ ⌷tan⌷ ⌷1⌷ ⌷.⌷ ⌷2⌷ = 50.2° (to 1 d.p.)

or your calculator may require

⌷1⌷ ⌷.⌷ ⌷2⌷ ⌷SHIFT⌷ ⌷tan⌷ = 50.2° (to 1 d.p.)

Here is an exam question ... ... and its solution

Work out the following. Give your answers to 2 d.p.

a) $4^2 + 3^2 - 2 \times 4 \times 3 \times \cos 20°$

b) $2.7 \times 10^{-6} \times 8.2 \times 10^{-4}$

c) $\sqrt{3} + 5 \cos 40°$ [4]

a) 2.45

b) 2.21×10^{-9}

c) 2.61

Now Try These Exam Questions

1 Find

a) $(3.6 \times 10^{-10})^{\frac{1}{2}}$

b) $(3.6 \times 10^{-9})^{\frac{1}{2}}$ [2]

2 Work out these.

a) $6.3 \times 10^9 + 5.8 \times 10^{10}$

b) $\dfrac{(9.52 \times 10^{14})^2}{8 \times 10^{-3}}$ [3]

More Exam Practice NE12

Expansions, factors and indices

Multiplying two brackets

- When expressions of the form $(a + b)(c + d)$ are expanded, every term in the first bracket is multiplied by every term in the second bracket.
  $$(a + b)(c + d) = ac + ad + bc + bd$$
- You may have various strategies for organising this work, for example 'FOIL' (firsts, outsides, insides, lasts) or making a multiplication table.
- Having expanded the brackets to four terms, you can sometimes combine two 'like terms' to simplify the expression.

> **Chief Examiner Says**
>
> 'Multiply out', 'Expand' and 'Remove the brackets' all mean the same thing.

Test Yourself (1)

Multiply out the brackets in the following.
a) $(x + 4)(x + 2)$
b) $(3a + 5)(4a - 3)$
c) $(x - 3)^2$

More Practice A19

Common factors

- Look for every number and letter that is common to every term and write these outside the bracket.
- Write the terms in the bracket needed to give the original expression.

> **Chief Examiner Says**
>
> Always try to take out as big a factor as possible. In part a) 2 is a common factor, 4 is a common factor, and x is a common factor. The biggest common factor is $4x$.

Test Yourself (2)

Factorise these.
a) $12x^2 - 8xy$
b) $6x^2y + 3xy^2$
c) $6ab^2 + 4b^2 - 2abc$

More Practice A20

Indices

These are used with both letters and numbers.
- As well as the rules for indices on page 17,
  $$a^{-n} = \frac{1}{a^n}$$
  $$a^{\frac{1}{n}} = \sqrt[n]{a}$$
  $$a^{\frac{m}{n}} = (\sqrt[n]{a})^m = \sqrt[n]{a^m}$$

More Practice A21

> **Solutions**
>
> **Test Yourself (1)**
> a) $x^2 + 2x + 4x + 8 = x^2 + 6x + 8$
> b) $12a^2 - 9a + 20a - 15 = 12a^2 + 11a - 15$
> c) $(x - 3)(x - 3) = x^2 - 3x - 3x + 9 = x^2 - 6x + 9$
>
> Multiply the bracket by itself
>
> **Test Yourself (2)**
> a) $4x(3x - 2y)$
> b) $3xy(2x + y)$
> c) $2b(3ab + 2b - ac)$

Here is an exam question ...   ... and its solution

a) Expand $(2q - 3)(q + 5)$. **[3]**

b) Factorise completely $6p^2 - 8p$. **[2]**

c) Simplify $\dfrac{2a^4 \times 4a^2}{a^3}$. **[2]**

d) Find the value of $\left(\dfrac{49}{4}\right)^{-\frac{3}{2}}$ as a fraction. **[2]**

a)

×	$2q$	-3
q	$2q^2$	$-3q$
5	$10q$	-15

Answer:
$2q^2 + 7q - 15$

b) $2p(3p - 4)$

c) $\dfrac{8a^6}{a^3} = 8a^3$

d) $\left(\dfrac{49}{4}\right)^{-\frac{3}{2}} = \left(\dfrac{4}{49}\right)^{\frac{3}{2}} = \left(\dfrac{2}{7}\right)^3 = \dfrac{8}{343}$

Invert   Square root   Cube

Now Try These Exam Questions

1 Factorise completely $12p^2q - 15pq^2$. **[2]**

2 Simplify these.

a) $3a^2b \times 4ab^3$ **[2]**

b) $\dfrac{a^3b^5}{a^2b^3}$ **[2]**

3 Simplify $\dfrac{m^3n^3 \times m^4n^2}{m^5n}$ **[3]**

4 Expand and simplify these.

a) $(3 - x)^2$ **[2]**

b) $(3x - 2)(x + 4)$ **[2]**

More Exam Practice AE9

Formulae

Rearranging formulae

- To change the subject of a formula, use the equation rule of doing the same to both sides to get the new subject on one side of the formula.
- See section 1 page 20 for simple examples.

Test Yourself (1)

Make u the subject of this formula.

$s = ut + \frac{1}{2}at^2$

More Practice A22

Solutions

Test Yourself (1)

$2s = at^2 + 2ut$ ← Multiply both sides by 2, to eliminate fractions

$2s - at^2 = 2ut$ ← Take at^2 from both sides

$\dfrac{2s - at^2}{2t} = u$ ← Divide both sides by 2t. The fraction line acts as a bracket

$u = \dfrac{2s - at^2}{2t}$ ← Rewrite with u on the left

Chief Examiner Says

Remember that in formulae you should use a fraction line for divide not a divide (÷) sign.

Powers of the subject

- If the subject is raised to a power, for example v^2, first make v^2 the subject and then find the square root of both sides.
- For cubes find the cube root, for power four the fourth root and so on.
- If the subject is in a square root, for example $\sqrt{a}$, first make $\sqrt{a}$ the subject and then square both sides.

Test Yourself (1)

Make u the subject of $v^2 = u^2 + 2as$.

More Practice A23

Subject twice in formula

If the required subject is in the formula twice, carry out the following steps:
- Rearrange so that all the terms involving the subject are on one side of the equation and all the other terms are on the other side.
- Take the subject out as a common factor.
- Divide both sides by the bracket.
- If necessary, rewrite with the subject on the left.

Test Yourself (2)

Make x the subject of $ax - by = cx + d$.

More Practice A24

Here is an exam question ...	... and its solution
Use the formula $F = 2(C^2 + 15)$ to find an expression for C in terms of F. **[3]**	$F = 2(C^2 + 15)$ $F = 2C^2 + 30$ $2C^2 = F - 30$ $C^2 = \frac{1}{2}(F - 30)$ $C = \sqrt{\frac{1}{2}(F - 30)}$

Now Try These Exam Questions

1. Rearrange the formula $P = 5\sqrt{V}$ to make V the subject. **[2]**

2. Given that $V = \frac{1}{3}\pi r^2 h$, express r in terms of V, h and π. **[2]**

3. Rearrange each of the following to give d in terms of e.
   a) $de = 5d + 3$ **[3]**
   b) $\dfrac{3d - 7}{4 + 5d} = e$ **[4]**

More Exam Practice AE10

Solutions

Test Yourself (1)

$v^2 - 2as = u^2$ ◄ Take $2as$ from both sides

$\sqrt{v^2 - 2as} = u$ ◄ Find the square root of both sides

$u = \sqrt{v^2 - 2as}$ ◄ Rewrite with u on the left

Test Yourself (2)

$ax = cx + d + by$ ◄ Add by to both sides

$ax - cx = d + by$ ◄ Subtract cx from both sides

$x(a - c) = d + by$ ◄ Take x out as a common factor

$x = \dfrac{d + by}{a - c}$ ◄ Divide both sides by $(a - c)$

Direct and inverse proportion

Direct proportion

- In direct proportion, both variables change in the same way – either both getting larger or both getting smaller.
- Using symbols for direct proportion, y is proportional to x is written as $y \propto x$ and y is proportional to x^2 is written as $y \propto x^2$.
- The formulae for these are $y = kx$ and $y = kx^2$.

When x and y vary in direct proportion, the graph of y against x is a straight line passing through the origin.

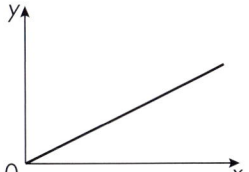

Chief Examiner Says

These are the most common direct proportions but you could also meet $y \propto x^3$ and $y \propto \sqrt{x}$.

Test Yourself (1)

The value, £V, of a diamond is proportional to the square of its mass, W g. A diamond weighing 14 g is worth £490.

a) Find the value of a diamond weighing 40 g.

b) Find the mass of a diamond worth £6000.

More Practice A25

Inverse proportion

- In inverse proportion, when one variable increases the other variable decreases.
- Using symbols, y is inversely proportional to x is written as $y \propto \dfrac{1}{x}$.

  and y is inversely proportional to x^2 is written as $y \propto \dfrac{1}{x^2}$.

- The formulae for these are $y = \dfrac{k}{x}$ and $y = \dfrac{k}{x^2}$.

Chief Examiner Says

These are the most common proportions but you could also meet $y \propto \dfrac{1}{\sqrt{x}}$.

Solutions

Test Yourself (1)

Method 1

$V \propto W^2$ or $V = kW^2$

$490 = k \times 14^2$

$k = \dfrac{490}{14^2} = 2.5$

So $V = 2.5W^2$

a) $V = 2.5 \times 40^2$
$= £4000$

b) $6000 = 2.5W^2$
$W^2 = 6000 \div 2.5 = 2400$
$W = \sqrt{2400} = 48.99$ g or 49 g, to the nearest gram

Method 2

a) Scale factor for $V = \dfrac{40}{14}$

So scale factor for $W = \left(\dfrac{40}{14}\right)^2$

So $V = 490 \times \left(\dfrac{40}{14}\right)^2 = £4000$

b) Scale factor for $W = \dfrac{6000}{490}$

So scale factor for $V = \sqrt{\dfrac{6000}{490}}$

$W = 14 \times \sqrt{\dfrac{6000}{490}} = 48.99$ g

or 49 g, to the nearest gram

Test Yourself (1)

The volume, V m³, of a given gas is inversely proportional to the pressure P N/m².
When $V = 2$ m³, $P = 500$ N/m².

a) Find the volume when the pressure is 400 N/m².

b) Find the pressure when the volume is 5 m³.

More Practice A26

Here is an exam question ...   ... and its solution

From a point h metres above sea level the distance, d kilometres, to the horizon is given by $d \propto \sqrt{h}$.
When $h = 100$ m, $d = 35$ km.
Find d when $h = 25$ m.   **[3]**

$d \propto \sqrt{h}$ or $d = k\sqrt{h}$
$35 = k\sqrt{100}$
$k = 3.5$
So $d = 3.5 \times \sqrt{25}$
     $= 17.5$ km

Now Try These Exam Questions

1 A bullet fired from a gun is slowed down by air resistance. The resistance is proportional to the square of the speed. If the resistance is 100 N when the speed is 300 m/s, find:

   a) the resistance when the speed is 600 m/s.   **[2]**

   b) the speed if the resistance is 200 N. Give your answer to 3 s.f.   **[3]**

2 When x takes a certain value, the value of y is 10.
If this value of x is multiplied by 4, work out the value of y in each of the following cases.

   a) y is proportional to x **[1]**

   b) y is proportional to x^2 **[1]**

   c) y is inversely proportional to x **[1]**

3 y is inversely proportional to x^2 and $y = 9$ when $x = 2$.

   a) Find the equation connecting y and x.   **[3]**

   b) Use the equation to find the **values** of x when $y = 1$.   **[2]**

More Exam Practice AE11

Solutions

Test Yourself (1)

Method 1

$V \propto \dfrac{1}{P}$ or $V = \dfrac{k}{P}$

$2 = \dfrac{k}{500}$ so $k = 1000$

So $V = \dfrac{1000}{P}$

a) $V = \dfrac{1000}{400} = 2.5$ m³

b) $5 = \dfrac{1000}{P}$

   so $5P = 1000$

   so $P = \dfrac{1000}{5} = 200$ N/m²

Method 2

a) Scale factor for $P = \dfrac{400}{500} = \dfrac{4}{5}$

So scale factor for $V = \dfrac{5}{4}$

$V = 2 \times \dfrac{5}{4} = 2.5$ m³

b) Scale factor for $V = \dfrac{5}{2}$

So scale factor for $P = \dfrac{2}{5}$

So $P = 500 \times \dfrac{2}{5} = 200$ N/m²

Gradient and equations of straight lines

Gradient and *y*-intercept

- The gradient of a line is a number indicating how steep it is.
  The larger the number the steeper the line.

- Lines with positive gradient slope forwards /

  Lines with negative gradient slope backwards \

- Gradient = $\dfrac{\text{increase in } y}{\text{increase in } x}$.

- The *y*-intercept is the value of *y* where the line crosses the y-axis.

Test Yourself (1)

Find the gradient and y-intercept of these lines

a) 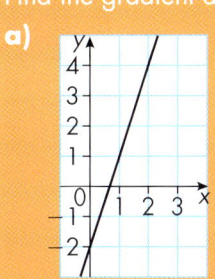   **b)**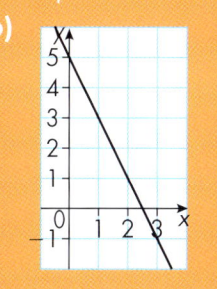

More Practice A27

The general equation of a straight line $y = mx + c$

- In the equation, *m* stands for the gradient of the line and *c* is the *y*-intercept.
- The equation of a line **must** be written in the form $y = mx + c$ for the two numbers to represent the gradient and the *y*-intercept.

Test Yourself (2)

a) Find the equations of these lines.

i) 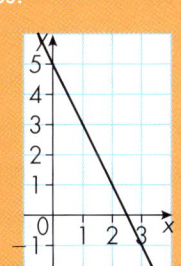   **ii)**

b) Find the gradient and y-intercept of these lines.
  i) $y = x + 4$ **ii)** $2y = 6x - 3$ **iii)** $5x - 2y = 12$

More Practice A28

Solutions

Test Yourself (1)

a) Gradient = $\frac{6}{2} = 3$
  y-intercept = -2

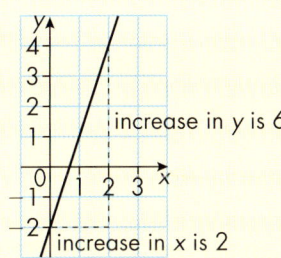

increase in y is 6

increase in x is 2

b) Gradient = $-\frac{4}{2} = -2$
  y-intercept = 5

Test Yourself (2)

a) i) Gradient = 3, y-intercept = -2.
  Equation is $y = 3x - 2$

ii) Gradient = -2, y-intercept = 5.
  Equation is $y = -2x + 5$

a) i) $y = 1x + 4$ $m = 1$ $c = 4$
ii) $y = 3x - 1\frac{1}{2}$ ← Dividing both sides by 2
  $m = 3, c = -1\frac{1}{2}$
iii) $5x = 12 + 2y$
  $5x - 12 = 2y$
  $y = 2\frac{1}{2}x - 6$ Rearranging to make y the subject
  $m = 2\frac{1}{2}, c = -6$

Parallel and perpendicular lines

- Lines which are parallel have the same gradient
- If two lines with gradients m_1 and m_2 are perpendicular then $m_1 m_2 = -1$.
- If a line has gradient m, then a line perpendicular to it has gradient $-\dfrac{1}{m}$.

More Practice A29

Test Yourself (1)

a) Find the equation of the line parallel to $y = 2x - 5$ and passing through $(0, 4)$.

b) Find the equation of the line perpendicular to $y = 4x + 1$ which passes through the point $(8, 5)$.

Here is an exam question ...   ... and its solution

Find the gradient and the equation of the straight line in the diagram.

[3]

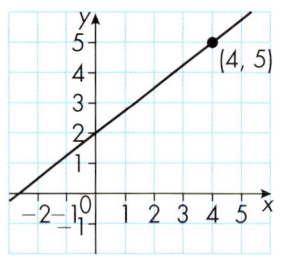

Gradient $= \dfrac{3}{4}$

Equation is $y = \dfrac{3}{4}x + 2$

Now Try These Exam Questions

1 **a)** Write down the gradient and y-intercept of the line with equation $y = 4 - 2x$. **[2]**

b) Write down the equation of the line parallel to $y = 4 - 2x$ and passing through the point $(0, -1)$. **[2]**

2 Find the equations of two straight lines which pass through the point $(1, 1)$, one parallel to $x + y = 1$ and the other perpendicular to $x + y = 1$. **[4]**

More Exam Practice AE12

Quadratic functions and equations

Factorising quadratics

- Some quadratic functions can be factorised.
- Quadratic functions of the form $x^2 + bx + c$ can be factorised to the form $(x + p)(x + q)$ where $p + q = b$ and $p \times q = c$. If c is positive then p and q are both positive or both negative. If c is negative then p and q are of opposite sign.
- Quadratic functions of the form $ax^2 + bx + c$ can be factorised to the form $(px + q)(rx + s)$ where $p \times r = a$, $q \times s = c$ and $ps + qr = b$. If c is positive then q and s are both positive or both negative. If c is negative then q and r are of opposite sign.

Solutions

Test Yourself (1)

a) From $y = 2x - 5$, $m = 2$
From $(0, 4)$, $c = 4$
Equation is $y = 2x + 4$

b) $m_1 = 4$ so $m_2 = -\dfrac{1}{4}$.

Equation is $y = -\dfrac{1}{4}x + c$

Since line goes through $(8, 5)$, substituting gives $5 = \left(-\dfrac{1}{4}\right) \times 8 + c$

$5 = -2 + c$ so $c = 7$

So equation is $y = -\dfrac{1}{4}x + 7$

Test Yourself (1)

Factorise these.

a) $x^2 + x - 6$ b) $3x^2 - 13x + 4$

More Practice A30

Difference of two squares

- Expressions of the form $a^2 - b^2$ can be factorised in the form $(a + b)(a - b)$.

Test Yourself (2)

Factorise these.

a) $x^2 - 16$

b) $2a^2 - 50b^2$

More Practice A31

Solving quadratic equations by factorising

- If the product of two numbers is 0 then one of the numbers must be zero.
- If a quadratic function equals 0 then one or other of its factors equals zero. Equating both factors to zero gives the two solutions to the equation.

Test Yourself (3)

Solve these.

a) $(3x - 2)(x + 1) = 0$

b) $x^2 - 3x = 0$

c) $x^2 + x = 6$

More Practice A32

Solutions

Test Yourself (1)

a) Look for two numbers which multiply to give -6 and add to give $+1$. They are -2 and 3.
Factors are $(x - 2)(x + 3)$

b) The only factors that give 3 are 3 and 1. Since c is positive and b is negative, q and s must both be both negative.
Factors are $(3x - ...)(x - ...)$
Possibilities for r and s are -4 and -1 or -1 and -4 or -2 and -2.
Checking the middle term shows that the correct factors are $(3x - 1)(x - 4)$ as this is the only way to get $-13x$ as the middle term.

Test Yourself (2)

a) $x^2 - 16 = x^2 - 4^2 = (x + 4)(x - 4)$

b) $2a^2 - 50b^2 = 2(a^2 - 25b^2)$
$= 2[a^2 - (5b)^2]$
$= 2(a + 5b)(a - 5b)$

> Look for a common factor

> Rearrange so that 0 is on one side of the equation

> Factorise

Test Yourself (3)

a) $3x - 2 = 0$ or $x + 1 = 0$
$3x = 2$ or $x = -1$
$x = \frac{2}{3}$ or $x = -1$

b) $x(x - 3) = 0$
$x = 0$ or $x - 3 = 0$
$x = 0$ or 3

c) $x^2 + x - 6 = 0$
$(x + 3)(x - 2) = 0$
$x + 3 = 0$ or $x - 2 = 0$
$x = -3$ or 2

Completing the square

- Functions of the form $x^2 + bx + c$ can be expressed in the form $(x + m)^2 - m^2 + c$ where $m = \frac{1}{2}b$.

More Practice A33

Solving quadratic equations that do not factorise

1 Completing the square

- Make sure the equation has 0 on the right-hand side.
- If the coefficient of x^2 is not 1, divide through the equation by the coefficient of x.
- Complete the square on the left-hand side.
- Rearrange the equation to the form $(x + m)^2 = p$.
- Take the square root of both sides, remembering $\pm$.

2 Formula

- The solutions to the equation $ax^2 + bx + c = 0$ are given by the formula

$$x = \frac{-b \pm \sqrt{b^2 - 4ac}}{2a}$$

- This formula is given on the examination paper. Take care with signs and always work out the numerator before dividing by $2a$.

More Practice A34

Here is an exam question ...   ... and its solution

a) Write $x^2 + 6x + 2$ in the form $(x + a)^2 + b$. **[2]**

b) Hence state the minimum value of y on the curve $y = x^2 + 6x + 2$. **[2]**

c) Solve the equation $x^2 + 6x + 2 = 0$. **[3]**

a) $x^2 + 6x + 2 = (x + 3)^2 - 9 + 2$
$\qquad\qquad\quad = (x + 3)^2 - 7$

b) The least value of $(x + 3)^2$ is 0, so the least value of y is -7.

c) $x^2 + 6x + 2 = 0$
$(x + 3)^2 - 7 = 0$
$(x + 3)^2 = 7$
$x + 3 = \pm\sqrt{7}$
$x = -3 \pm \sqrt{7}$
$\quad = -0.35 \text{ or } -5.65 \text{ to 2 d.p.}$

Solutions

Test Yourself (1)

$a = 10 \div 2 = 5$
$x^2 + 10x + 12 = (x + 5)^2 - 5^2 + 12 = (x + 5)^2 - 13$

Test Yourself (2)

$x^2 - 6x + 0.5 = 0$ ◄ Divide both sides by 2
$(x - 3)^2 - 3^2 + 0.5 = 0$
$(x - 3)^2 - 8.5 = 0$ ◄ Complete the square
$(x - 3)^2 = 8.5$
$(x - 3) = \pm\sqrt{8.5}$
$x = 3 \pm \sqrt{8.5}$
$\quad = 5.92 \text{ or } 0.08 \text{ correct to 2 decimal places.}$

Test Yourself (3)

$a = 2, b = -7, c = -3$

$$x = \frac{-(-7) \pm \sqrt{(-7)^2 - 4 \times 2 \times (-3)}}{2 \times 2}$$

$$= \frac{7 \pm \sqrt{49 + 24}}{4}$$

$$= \frac{7 + \sqrt{73}}{4} \text{ or } \frac{7 - \sqrt{73}}{4}$$

Press '=' before dividing by 4

$= 3.89 \text{ or } -0.39 \text{ correct to 2 decimal places.}$

1 a) Factorise completely $5x^2 - 20$. **[2]**

   b) i) Factorise $x^2 - 9x + 8$. **[2]**

      ii) Hence solve $x^2 - 9x + 8 = 0$. **[1]**

2 Solve the equation $2x^2 - 38x + 45 = 0$.
Give your answers to 2 decimal places. **[3]**

3 a) Write $x^2 - 12x + 2$ in the form
$(x - a)^2 - b$. **[2]**

   b) Hence find the minimum value of
$x^2 - 12x + 2$. **[1]**

   c) Solve the equation $x^2 - 12x + 2 = 0$. **[3]**

More Exam Practice AE13

Algebra

Simultaneous equations

- When solving simultaneous equations you are looking for a single value of x and a single value of y which satisfy both equations.

Elimination method

- This is the best method if both equations are in the form $ax + by = c$.
- Step 1. Make the coefficients of x or y equal by multiplying one or both of the equations by a numerical factor. Don't forget to multiply **every** term.
- Step 2. Eliminate the term with equal coefficients:
  - If the signs on the 'equalised' terms are the same, subtract the equations.
  - If the signs on the 'equalised' terms are different, add the equations.

  This gives the solution for one of the variables.
- Step 3. Substitute this value into one of the equations to find the value of the other variable.
- Step 4. Write down both values as your answer.
- You can check your answer by substituting both values back into the equation **not** used in Step 3.

Test Yourself (1)

Solve the following.

a) $2x + 3y = 6$ (1)
    $x + y = 4$ (2)

b) $3x - 5y = 1$ (1)
    $2x + 3y = 7$ (2)

Solutions

Test Yourself (1)

Multiply (2) by 2 to make the x terms the same

a) $2x + 3y = 6$
$2x + 2y = 8$

$(1) - (2)$ $y = -2$

The signs on the 2x terms are the same so subtract the equations

$x + (-2') = 4$

     $x = 6$

Substitute in (2) as it is easier

Solution $x = 6$, $y = -2$

Check in (1)

$2x + 3y = 2 \times 6 + 3 \times (-2) = 12 - 6 = 6$ ✓

This time both equations need to be multiplied. Make the y terms the same. (1) × 3 and (2) × 5

b) $9x - 15y = 3$
$10x + 15y = 35$

$(1) + (2)$ $19x = 38$
           $x = 2$

Signs in the y terms are different so add

$2 \times 2 + 3y = 7$
       $3y = 3$
        $y = 1$

Substitute in (2)

Solution is $x = 2$, $y = 1$

Check in (1) $3 \times 2 - 5 \times 1 = 1$ ✓

Substitution method

- This is the best method if y is the subject of one or both of the equations.
- Step 1. Substitute the equation with y as the subject into the other equation.
- Step 2. Solve this equation to find the value of x.
- Step 3. Substitute this value into the equation with y as the subject to find the value of y.
- If x is the subject of one of the equations the method can be reversed by substituting for x.

Test Yourself (1)

Use the substitution method to solve.

$2x + 3y = 10$ (1)
$y = 2x + 6$ (2)

More Practice A35

Simultaneous equations: one linear, one quadratic

- If both equations are in the form '$y =$ ', then eliminate y by equating them. Otherwise rearrange the linear equation so that y is the subject, then substitute for y in the quadratic equation.
- Simplify and rearrange the resulting quadratic equation so that one side is zero.
- Solve the quadratic equation using the methods in the previous section.
- Substitute each value of x in the linear equation to find the corresponding value for y.

Test Yourself (2)

Find the coordinates of the intersection of the line $y = 3x + 1$ and the curve $y = x^2 + 4x - 1$.

Chief Examiner Says

If you are asked to solve two simultaneous equations give the solution as two ordered pairs, each x value with the corresponding y value. If you are asked to find the intersection of two graphs, give the answers as coordinates.

More Practice A36

Simultaneous equations: one linear, one a circle

- The equation of a circle, centre the origin, with radius r is $x^2 + y^2 = r^2$. You only have to deal with circles where the centre is the origin.
- Make x or y the subject of the linear equation, whichever is the easier.
- Substitute this for x or y in the equation of the circle.
- Solve the quadratic equation using the methods in the previous section.
- Substitute each value in the linear equation to find the corresponding value of the other variable.

Solutions

Test Yourself (1)

$2x + 3(2x + 6) = 10$ Substitute (2) in (1). Don't forget the brackets
$2x + 6x + 18 = 10$
$8x = -8$
$x = -1$
$y = 2 \times (-1) + 6 = 4$ Substitute $x = -1$ in (2)
Solution $x = -1$, $y = 4$

Test Yourself (2)

$x^2 + 4x - 1 = 3x + 1$
$x^2 + x - 2 = 0$
$(x + 2)(x - 1) = 0$
So $x + 2 = 0$ or $x - 1 = 0$
$x = -2$ or $x = 1$
When $x = -2$, $y = 3 \times (-2) + 1 = -6 + 1 = -5$
When $x = 1$, $y = 3 \times 1 + 1 = 4$
So the graphs intersect at $(-2, -5)$ and $(1, 4)$.

Test Yourself (1)

Find the coordinates of the points of intersection of the straight line $x + 2y = 1$ and the circle $x^2 + y^2 = 25$.

More Practice A37

Here is an exam question ...

a) The line $y = 3x - 2$ intersects the circle $x^2 + y^2 = 12$. Show that $5x^2 - 6x - 4 = 0$ at the points of intersection. **[2]**

b) Hence find the coordinates of the points where they intersect, giving your answers to 2 decimal places. **[3]**

... and its solution

a) Substituting for y into the equation of the circle,
$$x^2 + (3x - 2)^2 = 12$$
$$x^2 + 9x^2 - 12x + 4 = 12$$
$$10x^2 - 12x - 8 = 0 \quad \text{◄ Divide by 2}$$
$$5x^2 - 6x - 4 = 0$$

b) Using the quadratic formula,
$a = 5$, $b = -6$, $c = -4$
$$x = \frac{6 \pm \sqrt{36 - 4 \times 5 \times (-4)}}{2 \times 5}$$

$$= -0.477... \text{ or } 1.677...$$

Substitute for x in $y = 3x - 2$

The line $y = 3x - 2$ and the circle $x^2 + y^2 = 12$ intersect at $(-0.48, -3.43)$ and $(1.68, 3.03)$.

Now Try These Exam Questions

1 Solve algebraically these pairs of simultaneous equations.

   a) $5x + 4y = 13$
   $3x + 8y = 5$ **[3]**

   b) $4x + 3y = 5$
   $2x + y = 1$ **[3]**

   c) $2x - 3y = 9$
   $5x + 2y = -25$ **[4]**

2 Find algebraically the intersections of the line $y = 3x - 1$ and the circle $x^2 + y^2 = 12$. Give the coordinates to 1 d.p. **[7]**

More Exam Practice AE14

Solutions

Test Yourself (1)

$x + 2y = 1$
$x = 1 - 2y$ ◄ Make x the subject of the linear equation
$(1 - 2y)^2 + y^2 = 25$ ◄ Substitute in the circle equation. Don't forget the brackets
$1 - 4y + 4y^2 + y^2 = 25$
$5y^2 - 4y - 24 = 0$ ◄ Multiply out the bracket

This does not factorise so use the quadratic formula.
$a = 5$, $b = -4$, $c = -24$

$$y = \frac{-(-4) \pm \sqrt{(-4)^2 - 4 \times 5 \times (-24)}}{2 \times 5}$$

$= -1.827...$ or $2.627...$
When $y = 2.627$, $x = 1 - 2 \times 2.627 = -4.254$
When $y = -1.827$, $x = 1 - 2 \times (-1.827) = 4.654$
So correct to 2 decimal places the points of intersection are $(-4.25, 2.63)$ and $(4.65, -1.83)$.

Algebraic fractions

Simplifying fractions

- The rules for simplifying algebraic fractions are the same as for numerical fractions.

Chief Examiner Says

Only cancel common factors. An error frequently seen is to cancel the x^2 in expressions such as $\dfrac{x^2 + 3x + 2}{x^2 - 1}$. You cannot cancel x^2 because it is not a common factor.

Test Yourself (1)

Simplify the following.

a) $\dfrac{x^2 + 3x + 2}{x^2 - 1}$

b) $\dfrac{2x + 1}{3} - \dfrac{x + 2}{5}$

c) $\dfrac{1}{x - 1} + \dfrac{1}{x + 2}$

More Practice A38

Solving equations with fractions

- Multiply both sides by the common denominator.

Test Yourself (2)

Solve these.

a) $\dfrac{x + 3}{2} + \dfrac{x + 2}{3} = 4$

> This could be written as $\frac{1}{2}(x + 3) + \frac{1}{3}(x + 2) = 4$

b) $\dfrac{3}{x - 2} - \dfrac{1}{x + 1} = 1$

More Practice A39

Solutions

Test Yourself (1)

a) $\dfrac{(x + 1)(x + 2)}{(x + 1)(x - 1)}$ ← Factorise the quadratics

$= \dfrac{x + 2}{x - 1}$ ← Cancel out the common factor

b) $\dfrac{5(2x + 1)}{15} - \dfrac{3(x + 2)}{15}$ ← Common denominator is 15. Multiply numerator and denominator by the appropriate factor

$= \dfrac{10x + 5 - 3x - 6}{15}$ ← Combine and multiply out the brackets. Don't forget that $(-3) \times 2 = -6$

$= \dfrac{7x - 1}{15}$

c) $\dfrac{x + 2}{(x - 1)(x + 2)} + \dfrac{x - 1}{(x - 1)(x + 2)}$

$= \dfrac{x + 2 + x - 1}{(x - 1)(x + 2)}$ ← Common denominator is $(x - 1)(x + 2)$. Multiply numerator and denominator by the appropriate factor

$= \dfrac{2x + 1}{(x - 1)(x + 2)}$ ← Combine

Test Yourself (2)

a) $\dfrac{\overset{3}{\cancel{6}}(x + 3)}{\cancel{2}} + \dfrac{\overset{2}{\cancel{6}}(x + 2)}{\cancel{3}} = 24$ ← Multiply both sides by 6 and cancel common factors

$3x + 9 + 2x + 4 = 24$ ← Multiply out the brackets

$5x + 13 = 24$

$5x = 11$

$x = 2\frac{1}{5}$

b) $\dfrac{3\cancel{(x - 2)}(x + 1)}{\cancel{x - 2}} - \dfrac{(x - 2)\cancel{(x + 1)}}{\cancel{x + 1}} = (x - 2)(x + 1)$ ← Multiply both sides by $(x - 2)(x + 1)$ and cancel common factors

$3x + 3 - x + 2 = x^2 - x - 2$

$0 = x^2 - 3x - 7$

$x = \dfrac{3 \pm \sqrt{9 + 28}}{2} = 4.54 \text{ or } -1.54$

Here is an exam question ...

Michael drives 70 miles to work at an average speed of v miles per hour.

On the return journey he travels 5 miles per hour faster and takes $\frac{1}{4}$ hour less.

a) i) Write down expressions in v for the two journey times. **[2]**

   ii) Hence form an equation in v and show that it simplifies to $v^2 + 5v - 1400 = 0$. **[3]**

b) Solve the equation to find v. **[2]**

... and its solution

a) i) $\dfrac{70}{v}$ and $\dfrac{70}{v+5}$

  ii) $\dfrac{70}{v} - \dfrac{70}{v+5} = \dfrac{1}{4}$

$$70(v+5) - 70v = \frac{1}{4}v(v+5)$$

$$1400 = v^2 + 5v$$

$$v^2 + 5v - 1400 = 0$$

b) $(v - 35)(v + 40) = 0$

$v = 35$

($v = -40$ is not possible.)

Now Try These Exam Questions

1 a) Simplify this expression. $\dfrac{x^2 - 9}{x^2 - x - 6}$ **[3]**

  b) Use algebra to solve this equation.

$$\frac{12}{3x+1} - \frac{5}{x+1} = 1 \qquad \textbf{[7]}$$

2 Simplify these.

  a) $\dfrac{2}{x-3} - \dfrac{1}{x}$ **[3]**

  b) $\dfrac{2}{x+1} + \dfrac{3}{x-2}$ **[4]**

3 Jane took part in a sponsored cycle ride.
She cycled from her home town to Blackpool and back.
The distance from her home town to Blackpool is 48 km.
Jane cycled to Blackpool at an average speed of 12 km/h.

  a) Find the time she took to cycle to Blackpool. **[2]**

Her average speed for the return journey was 8 km/h.

  b) Calculate her average speed for the **whole** journey. **[4]**

Sachin lives next door to Jane and took part in the same sponsored cycle ride.
He cycled to Blackpool at an average speed of x km/h.
He returned at an average speed of 5 km/h slower than on the outward journey.
The total journey took him 8 hours.

  c) i) Write down an equation in x and show that it simplifies to

$$\frac{6}{x} + \frac{6}{x-5} = 1 \qquad \textbf{[2]}$$

    ii) Use algebra to find the solutions to the equation in part **i)**. **[6]**

   iii) Hence find Sachin's average speed for the journey from Blackpool to his home. **[1]**

More Exam Practice AE15

Graphical solution of simultaneous equations and linear inequalities

Graphical solution of simultaneous equations

- Make a table of values for each equation and plot the graphs on the same pair of axes.
- Write down the coordinates of the point(s) where the graphs cross.

Test Yourself (1)
Solve graphically $y = 4x - 1$ and $x + y = 4$.

Test Yourself (2)
Solve graphically
$y = x + 1$
$x^2 + y^2 = 4$

More Practice A40

Graphical solution of a set of linear inequalities

- Step 1. Draw the boundary lines whose equations are the inequalities with the inequality signs replaced by equal signs. If the inequality sign is $\leq$ or $\geq$, use a full line. If the inequality sign is $<$ or $>$ then use a dotted line.
- Step 2. Decide which side of each line represents the given inequality. If the line does not go through the point (0, 0), this is the best point to use as a test. If the line does go through the point (0, 0) then use another point such as (1, 0).
- Step 3. Indicate clearly the **single** region which satisfies all of the inequalities. If there is more than one inequality it is probably best to use the convention to shade the **unwanted** side of the line as this leaves the part of the graph left unshaded as the required solution set.

Solutions

Test Yourself (1)

$y = 4x - 1$

x	0	1	2
y	−1	3	7

$x + y = 4$

x	0	4	1
y	4	0	3

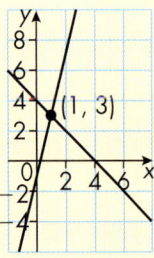

Answer: $x = 1$, $y = 3$

Test Yourself (2)

x	−1	0	1
y = x + 1	0	1	2

It is better to draw a circle with a pair of compasses than to plot points.
Here the centre is (0, 0) and the radius is 2.

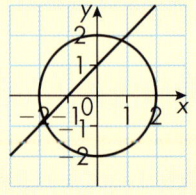

Don't forget to write your answer as pairs with each x value with its corresponding y value

The solutions are $x = 0.8$ and $y = 1.8$
or $x = -1.8$ and $y = -0.8$

Test Yourself (1)

Find the region satisfied by

$x \geqslant 1$ $x + y < 5$ $y \geqslant 2x - 6$

More Practice A41

Here is an exam question ... ... and its solution

Write down the three inequalities that are satisfied in the shaded region.

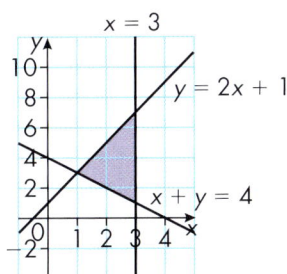

$x = 3$

$y = 2x + 1$

$x + y = 4$

[3]

Choose a point in the region, say $x = 2$, $y = 3$.

$y = 2x + 1$

$3 \leqslant 2 \times 2 + 1$ so $y \leqslant 2x + 1$

$x = 2$ $2 \leqslant 3$ so $x \leqslant 3$

$x + y = 4$ $2 + 3 \geqslant 4$ so $x + y \geqslant 4$

So inequalities are

$y \leqslant 2x + 1$, $x \leqslant 3$, $x + y \geqslant 4$

Now Try These Exam Questions

1 Solve these simultaneous equations graphically.

$y = 3x + 4$ $x + y = 2$ **[4]**

2 Find the region satisfied by the inequalities.

$x + y \leqslant 5$ $y \leqslant 2x - 1$ $y \geqslant 0$ **[5]**

3 Look at the graph. Write down the three inequalities which are satisfied in the shaded region. **[3]**

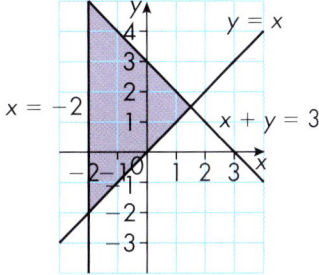

$x = -2$

$y = x$

$x + y = 3$

More Exam Practice AE16

Solutions

Test Yourself (1)

$x = 1$ is the line through (1, 0) and parallel to the y-axis.

$x + y = 5$ passes through (0, 5) and (5, 0) (dotted line since = not included).

$y = 2x - 6$ passes through (0, −6) and (2, −2) and (4, 2).

$x \geqslant 1$ For (0, 0), $x \geqslant 1$ is false. Shade the origin (unwanted) side.

$x + y > 5$ For (0, 0), $x + y < 5$ is true. Shade the opposite (unwanted) side to the origin.

$y \geqslant 2x - 6$ For (0, 0), $y \geqslant 2x - 6$ is true. Shade the opposite (unwanted) side to the origin.

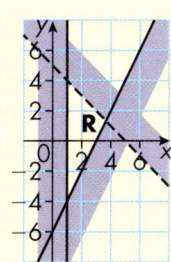

The region left unshaded is the required solution. It may make it clearer if an R is placed in this region.

Graphs of functions

Cubic graphs

● The cubic graph of $y = ax^3$ has this shape.

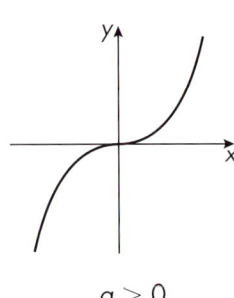
$a > 0$

or

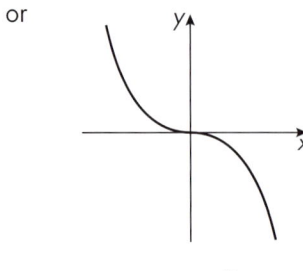
$a < 0$

● The general cubic function $ax^3 + bx^2 + cx + d$ has this shape.

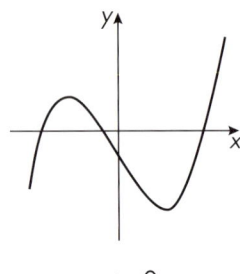
$a > 0$

or

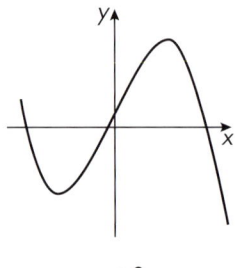
$a < 0$

● As with all graphs, the first step in plotting cubic graphs is to make a table of values.

Test Yourself (1)

a) Draw the graph of $y = x^3 - 3x^2$ for values of x from -2 to 4.

b) Use your graph to solve the equation $x^3 - 3x^2 = -1$.

More Practice A42

Solutions

Test Yourself (1)

a)

x	−2	−1	0	1	2	3	4
y	−20	−4	0	−2	−4	0	16

You may wish to put some extra rows in the table to help you with these calculations

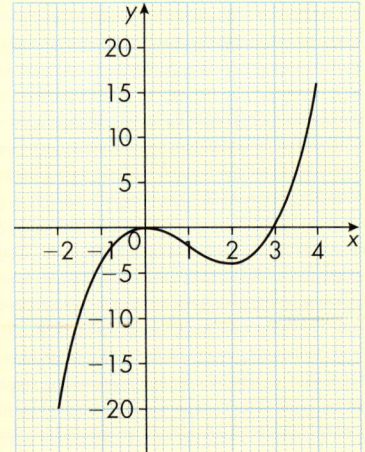

b) The solution is where the curve crosses the line $y = -1$.
The solution is $x = -0.5, 0.7$ or 2.9

Reciprocal and exponential graphs

- The reciprocal graph $y = \dfrac{a}{x}$ has this shape

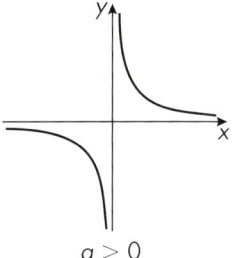

$a > 0$

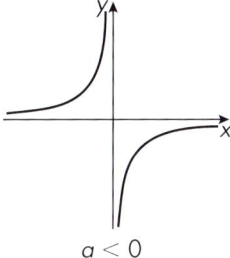

$a < 0$

> These graphs are in two separate curves. You cannot use 0 as a value for x, since you cannot divide a number by 0

- The exponential graph $y = a^x$ has this shape

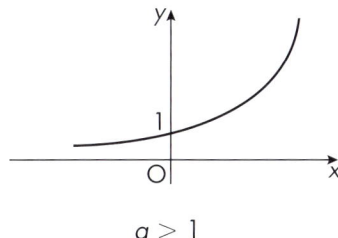

$a > 1$

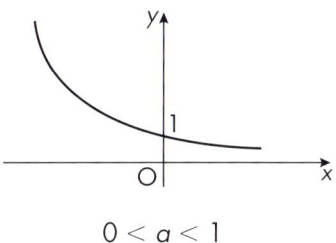

$0 < a < 1$

- You will only be asked to work out the y values for integer values of x.
- All exponential graphs go through the point $(0, 1)$ since $a^0 = 1$ for all positive values of a.
  When $a > 1$ they increase steeply for $x > 0$ and are small when x is negative.
  This is reversed for $0 < a < 1$.

Test Yourself (1)

Draw the graphs for $y = 3^x$ and $y = \left(\frac{1}{2}\right)^x$ on the same grid.
Use values of x from -3 to $+3$.

More Practice A43

Solutions

Test Yourself (1)

x	-3	-2	-1	0	1	2	3
$y = 3^x$	0.04	0.11	0.33	1	3	9	27

x	-3	-2	-1	0	1	2	3
$y = \left(\frac{1}{2}\right)^x$	8	4	2	1	0.5	0.25	0.125

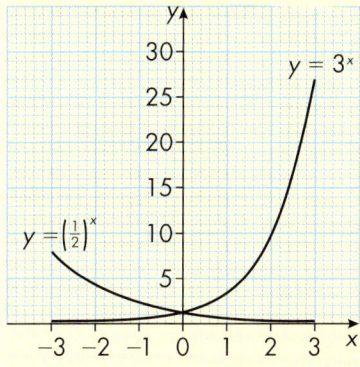

Trigonometrical graphs

- You need to know the shape of the graphs of $y = \sin x$ and $y = \cos x$
- They are both repeating wave shapes. They repeat every $360°$. The length of a repeating pattern like this is called the period of the graph
- The height of the wave above the mean is 1 unit. This height is called the amplitude of the wave
- The symmetries of these graphs may be used to find other angles which have the same sin or cos value. For instance
  $\sin 50° = \sin 130° = \sin 410°$ and $\cos 100° = -\cos 80°$

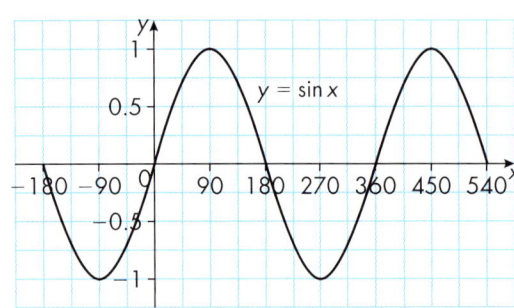

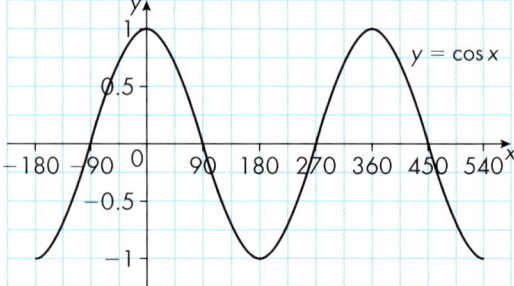

Test Yourself (1)

a) Use the graphs above to estimate the value of these.
 i) $\sin 120°$ **ii)** $\cos 200°$

b) Use the graphs above to find four approximate solutions to the following.
 i) $\sin x = 0.25$ **ii)** $\cos x = 0.75$

More Practice A44

Here is an exam question ...

a) Complete the table below for $y = x^3 - 2x^2 + 1$.

x	−1	−0.5	0
y		0.375	

[1]

b) Part of the graph is drawn on the grid. Add the three points from the table and complete the curve. [2]

c) Use the graph to solve the equation $x^3 - 2x^2 + 1 = 0$. [2]

... and its solution

a)

x	−1	−0.5	0
y	−2	0.375	1

b)

The dashed line shows the part of the graph that was given already

c) $x = -0.6$, 1 or 1.6.

The solution is where the graph crosses the line $y = 0$

Solutions

Test Yourself (1)

a) i) 0.9   **b) i)** 15°, 165°, 375°, 525°
 ii) −0.9   **ii)** −40°, 40°, 320°, 400°

Now Try These Exam Questions

1 a) Draw the graph of $y = \dfrac{2}{x}$ for values of x from -5 to 5. **[3]**

b) Use your graph to solve the equation $\dfrac{2}{x} = 0.8$. **[1]**

2 a) Complete the table of values and draw the graph of $y = x^3 - 4x - 1$ for values of x from -3 to 3. **[1]**

x	-3	-2	-1	0	1	2	3
y	-16		2	-1	-4	-1	14

b) Use your graph to solve the equation $x^3 - 4x - 1 = 0$. **[3]**

c) By drawing a suitable straight line on your graph, solve the equation $x^3 - 6x - 3 = 0$. **[4]**

More Exam Practice AE17

Transformations and functions

Function notation

- If y is a function of x then it can be written $y = f(x)$.
- $f(4)$ means the value of the function when $x = 4$.

Test Yourself (1)

If $f(x) = x^2 - 3x + 2$, find the values of these.
a) $f(4)$ **b)** $f(-2)$

Translations

- The graph of $y = f(x) + a$ is the graph of $y = f(x)$ translated by $\begin{pmatrix} 0 \\ a \end{pmatrix}$.

- The graph of $y = f(x + a)$ is the graph of $y = f(x)$ translated by $\begin{pmatrix} -a \\ 0 \end{pmatrix}$.

Test Yourself (2)

The graph shows $y = x^2$.
On a copy of this graph, sketch $y = x^2 + 2$.

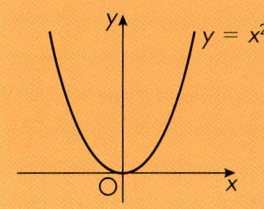

More Practice A45

Solutions

Test Yourself (1)

a) $f(4) = 4^2 - 3 \times 4 + 2 = 6$
b) $f(-2) = (-2)^2 - 3 \times (-2) + 2 = 12$

Test Yourself (2)

$y = x^2 + 2$ is the same shape as $y = x^2$ translated by $\begin{pmatrix} 0 \\ 2 \end{pmatrix}$.

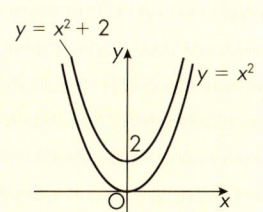

Reflections

- The graph of $y = f(-x)$ is the graph of $y = f(x)$ reflected in the y-axis.
- The graph of $y = -f(x)$ is the graph of $y = f(x)$ reflected in the x-axis.

Test Yourself (1)

The graph shows $y = x^2 - 3x$.

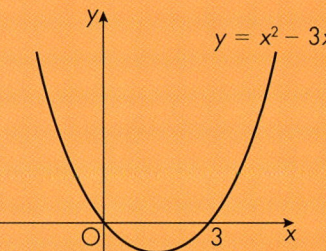

On a copy of the graph sketch the graph of $y = 3x - x^2$.

More Practice A46

Chief Examiner Says

There is no need to scale the axes for a sketch graph.
However, you should show important features, such as where
the graph crosses the axes.

Solutions

Test Yourself (1)

If $f(x) = x^2 - 3x$ then $-f(x) = 3x - x^2$
So the graph of $y = 3x - x^2$ is the reflection of
$y = x^2 - 3x$ in the x-axis.

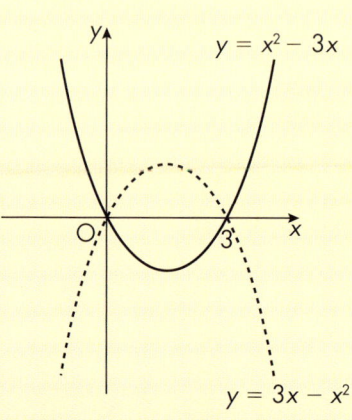

Stretches

- The graph of $y = af(x)$ is the graph of $y = f(x)$ stretched in the y direction with scale factor a.
- The graph of $y = f(ax)$ is the graph of $y = f(x)$ stretched in the x direction with scale factor $\dfrac{1}{a}$.

More Practice A47

Test Yourself (1)

The graph shows $y = \sin x$.

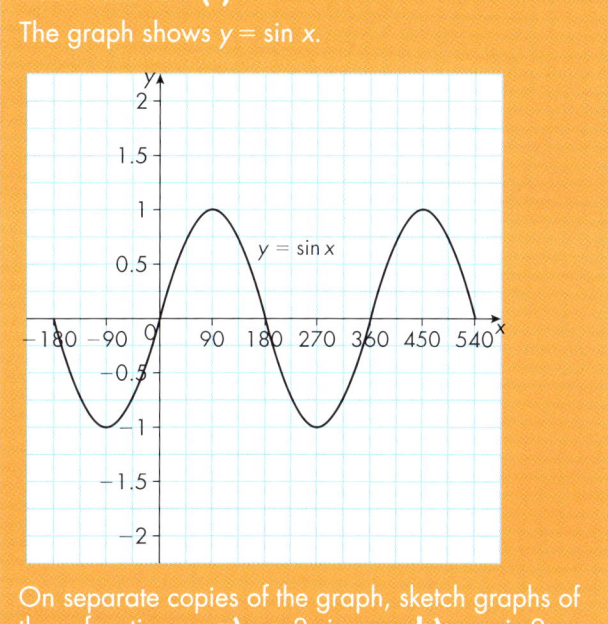

On separate copies of the graph, sketch graphs of these functions.   **a)** $y = 2 \sin x$ **b)** $y = \sin 2x$

Solutions

Test Yourself (1)

a) The graph of $y = 2 \sin x$ is the graph of $y = \sin x$ stretched in the y direction with scale factor 2.

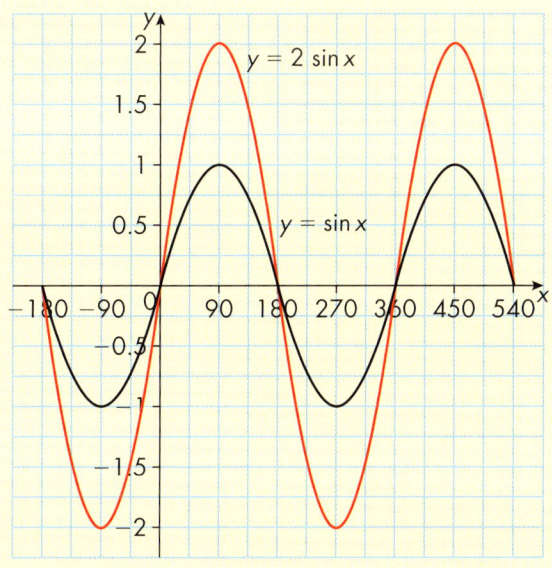

b) The graph of $y = \sin 2x$ is the graph of $y = \sin x$ stretched in the x direction with scale factor $\frac{1}{2}$.

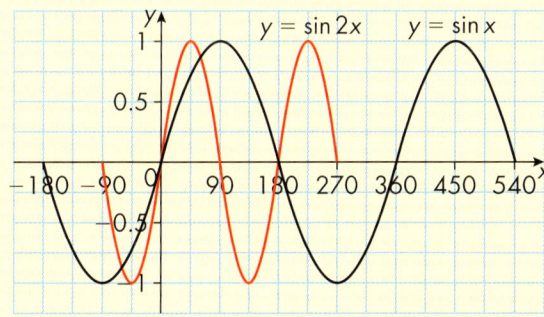

Here is an exam question ...

The sketch shows the graph of $y = f(x)$.

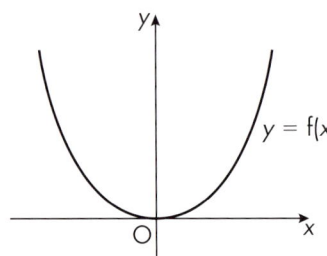

Sketch these graphs.

a) $y = 3 + f(x)$ **[1]**

b) $y = -\frac{1}{2}f(x)$ **[1]**

... and its solution

a)

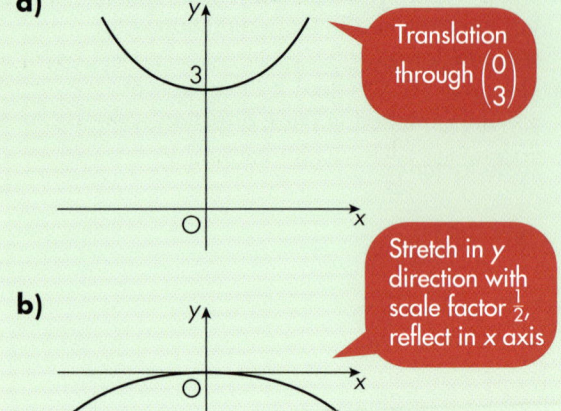

Translation through $\binom{0}{3}$

Stretch in y direction with scale factor $\frac{1}{2}$, reflect in x axis

b)

Now Try These Exam Questions

1 a) Sketch the graph of $y = x^2$.   **[1]**

   b) On the same diagram, sketch these graphs. In each case, describe the transformation.

     i) $y = -x^2$ **iii)** $y = \frac{1}{2}x^2$

     ii) $y = x^2 - 5$ **iv)** $y = (2x)^2$ **[4]**

2 On the same diagram, sketch the following graphs, labelling each graph clearly.

   a) $y = \sin x$ **[2]**

   b) $y = 3 \sin x$ **[1]**

   c) $y = \sin (x + 90°)$ **[1]**

More Exam Practice AE18

Graphs in real life

Graphs in real situations

To interpret real-life graphs
- Look at the labels on the axes – they tell you what the graph is about
- Look whether the graph is a straight line or a curve
- The slope of the graph gives you the rate of change
- For distance–time graphs, the rate of change is the velocity

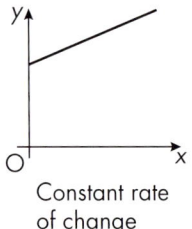

Constant rate of change

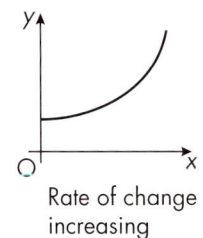

Rate of change increasing

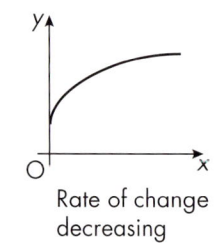

Rate of change decreasing

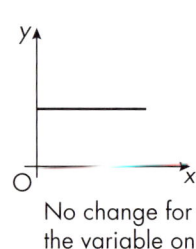

No change for the variable on the y-axis

Drawing graphs

- Always label your axis with the quantity and the unit.
- Sometimes only a sketch is asked for and not an accurate plot.
- Always check whether the rate of change is constant, increasing or decreasing.

 More Practice A48

Test Yourself (1)

Water is poured into this vessel at a constant rate.
Sketch a graph of depth of water (*d* cm) against time (*t* secs).

Here is an exam question ...

All these containers are full of liquid. The liquid runs out of each at a constant rate.
Match each graph to the correct container.

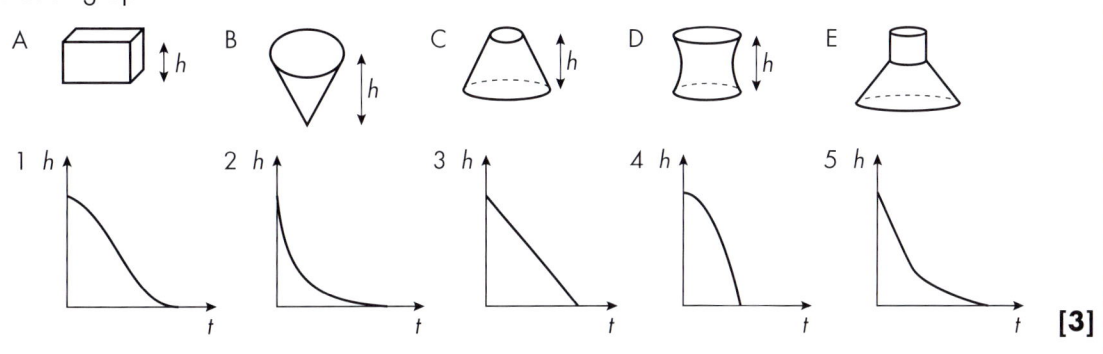

[3]

... and its solution

A ⟷ 3   B ⟷ 4   C ⟷ 2   D ⟷ 1   E ⟷ 5

Now Try These Exam Questions

1 Steve travelled from home to school by walking to a bus stop and then catching a school bus.

a) Use the information below to construct a distance–time graph for Steve's journey.
Steve left home at 08.00.
He walked at 6 km/h for 10 minutes.
He then waited for 5 minutes before catching the bus.
The bus took him a further 8 km to school at a steady speed of 32 km/h. **[3]**

b) How far was Steve from home at 08.20? **[1]**

 More Exam Practice AE19

2 The graph below describes a real-life situation. Describe a possible situation that is occurring.

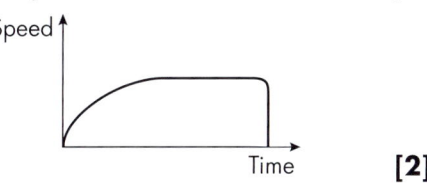

[2]

3 The diagrams show the cross-sections of three swimming pools. Water is pumped into all three at a constant rate. Sketch graphs of depth against time for each.

      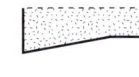

[3]

Solutions

Test Yourself (1)

Since the radius is decreasing, at first, the rate of change will increase.
At the top the radius is constant and so the rate of change will be constant.

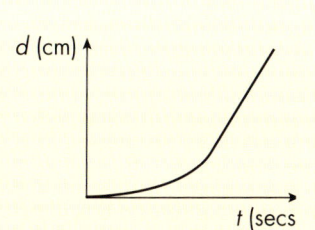

Properties of triangles and other shapes

Congruent triangles

- For any two triangles to be congruent they must each contain the same angles and each have sides of the same lengths.
- To prove two triangles are congruent there are four different conditions, one of which must be satisfied:

  1 SAS: Two sides and the included angle of one triangle are equal to two sides and the included angle of the other triangle.

  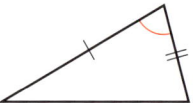 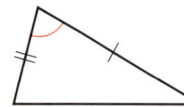

  2 SSS: The three sides of one triangle are equal to the three sides of the other triangle.

  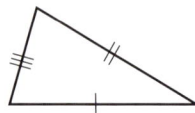 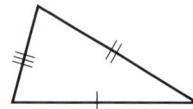

  3 ASA: Two angles and the side between them of one triangle are equal to two angles and the side between them of the other triangle.

  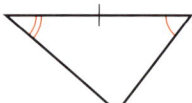 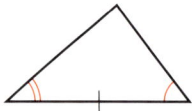

  4 RHS: Each triangle is right-angled and the hypotenuse and one other side of one triangle are equal to the hypotenuse and one other side of the other triangle.

  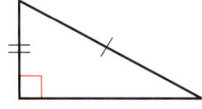 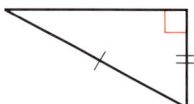

Test Yourself (1)

Are these triangles congruent?
If they are, use letters in the correct order to state clearly which triangles are equal. Give a reason for your answer. Also list the other sides and angles that are equal.

a)

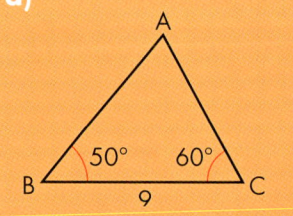 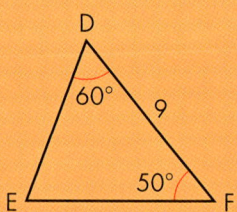

b)

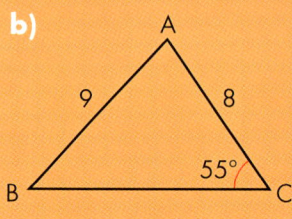 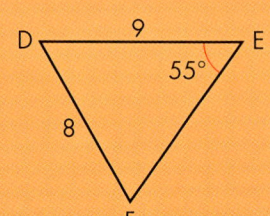

c)

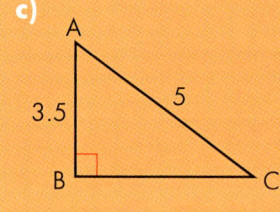 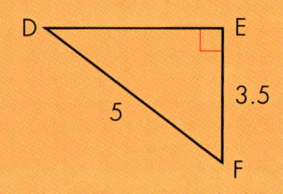

Chief Examiner Says

- It is important that the order of the letters is correct when stating whether two triangles are congruent.
- ≡ means 'is congruent to'.
- Often when finding equal lengths or angles you will need to give a reason why they are equal.

Solutions

Test Yourself (1)

a) Yes Triangle ABC ≡ Triangle EFD (ASA)
     Angle A = Angle E (third angle)
     AB = EF (opposite 60°)
     AC = ED (opposite 50°)

b) No The sides that are equal are not in the same corresponding position relative to the equal angle.

c) Yes Triangle ABC ≡ Triangle FED (RHS)
     Angle A = Angle F
     Angle C = Angle D
     BC = ED

Test Yourself (1)

Prove that the opposite sides of a parallelogram are equal.

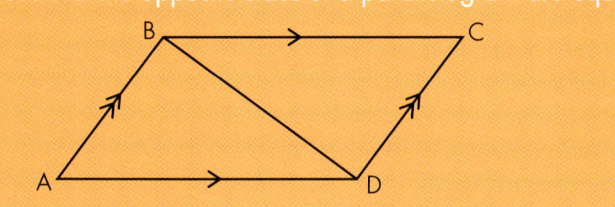

More Practice S23

Similar triangles

For triangles to be similar:
- All corresponding angles are equal.
- Corresponding sides have lengths in proportion. [That is one triangle is an enlargement of the other.]

More Practice S24

Test Yourself (2)

a) Prove that triangle ABC is similar to triangle PQR.

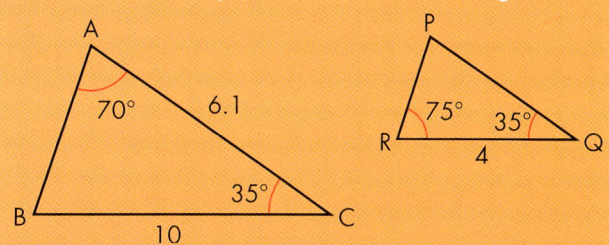

b) Find the length of PQ.

Here is an exam question ...   ... and its solution

Triangles AXB and CXD are similar.
a) Find the lengths of XC and DC.
b) Explain why AB is parallel to DC.

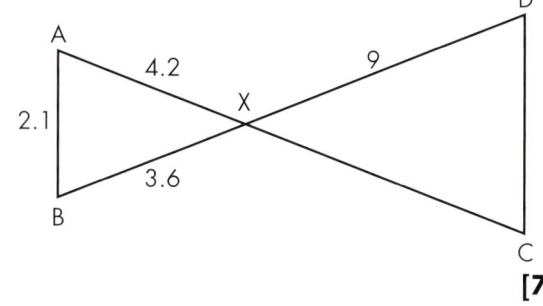

[7]

a) $\dfrac{XC}{4.2} = \dfrac{9}{3.6}$

$XC = \dfrac{9 \times 4.2}{3.6} = 10.5$ cm

$\dfrac{DC}{2.1} = \dfrac{9}{3.6}$

$DC = \dfrac{9 \times 2.1}{3.6} = 5.25$ cm

b) Angle BAX = angle DCX (corresponding angles in similar triangles). But these are alternate angles for AB and DC, making them parallel.

Solutions

Test Yourself (1)

In triangle ABD and triangle BDC:

Angle ABD = Angle BDC (alternate angles)
Angle ADB = Angle DBC (alternate angles)
Side DB is common to both triangles.
So triangle ABD ≡ triangle CDB (ASA).
Therefore the other corresponding sides are equal.
So AB = CD and AD = CB (opposite sides are equal).

Angle ABD means the angle at B formed by lines from A and D

Test Yourself (2)

a) Angle ABC = 180 − 70 − 35 = 75
Angle RPQ = 180 − 75 − 35 = 70
Triangles are similar as corresponding angles are equal.

b) $\dfrac{PQ}{4} = \dfrac{6.1}{10}$

PQ = 2.4 (1 d.p.)

Now Try This Exam Question

1 The diagram shows two squares, ABCX and PQRX.
Prove that angles PCX and RAX are equal. **[4]**

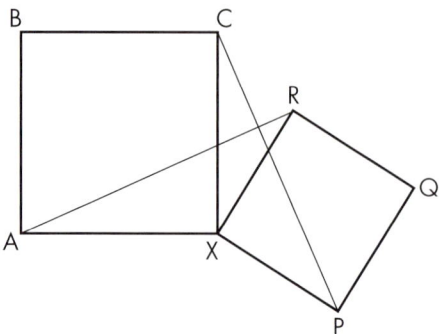

More Exam Practice SE6

Pythagoras and trigonometry

Pythagoras' theorem and trigonometry

● The three trigonometric ratios are:

$$\sin x = \frac{\text{Opposite}}{\text{Hypotenuse}} \qquad \cos x = \frac{\text{Adjacent}}{\text{Hypotenuse}} \qquad \tan x = \frac{\text{Opposite}}{\text{Adjacent}}$$

Test Yourself (1)

1 Find length x in this triangle.

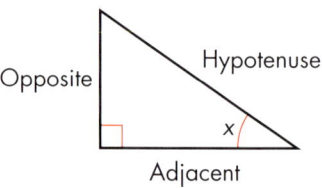

2 Find angle y in this triangle.

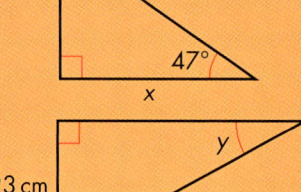

3 ABCDE is a square-based pyramid.
The edge of the base is 30 cm and the height, EH, is 42 cm.

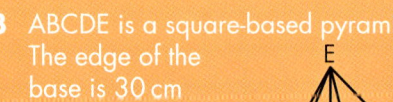
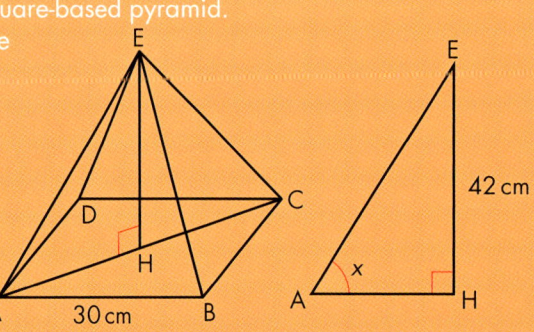

Calculate the following.

a) The length of AC   b) The angle of EAH
c) The length of EA

Chief Examiner Says

● When using trigonometry, label the sides with H, O and A. This will help you work out which formula to use.
● When finding angles remember to use the $\sin^{-1}$, $\cos^{-1}$ and $\tan^{-1}$ buttons. (This is often found by pressing SHIFT and sin, or 2nd F and sin.)
● In three-dimensional questions it is useful to draw out separately the right-angled triangle you are using.
● For accuracy, always use all the figures from one answer when you are using that value in another calculation.

More Practice S25

Solutions

Test Yourself (1)

1 $\cos 47° = \dfrac{x}{20}$

$x = 20 \cos 47°$

$x = 13.6$ cm

2 $\sin y = \frac{23}{40}$

$y = \sin^{-1}\left(\frac{23}{40}\right)$

$y = 35.1°$

3 a) $AC^2 = AB^2 + BC^2 = 900 + 900 = 1800$

$AC = \sqrt{1800} = 42.43$ cm

b) $\tan x = \dfrac{EH}{AH} = \dfrac{42}{21.21...} = 1.98$ angle EAH = 63.2°

c) By Pythagoras $EA^2 = EH^2 + AH^2$

$EA^2 = 42^2 + 21.21^2 = 2213.86$

$EA = 47.1$ cm

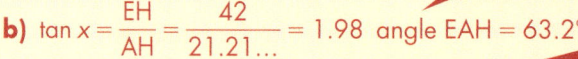

For accuracy, use all the figures from your answer for AC when finding angle EAH

If you kept all the figures in 21.21..., you should find that $21.21...^2 = 450$. Can you explain that?

Sine and cosine rules

For this triangle:

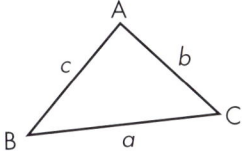

- The sine rule is
$$\frac{\sin A}{a} = \frac{\sin B}{b} = \frac{\sin C}{c}$$

- The cosine rule is
$$a^2 = b^2 + c^2 - 2bc \cos A$$
$$b^2 = c^2 + a^2 - 2ca \cos B$$
$$c^2 = a^2 + b^2 - 2ab \cos C$$

- The area of the triangle is
$$\text{Area} = \tfrac{1}{2}ab \sin C$$
$$= \tfrac{1}{2}bc \sin A$$
$$= \tfrac{1}{2}ac \sin B$$

More Practice S26

Test Yourself (1)

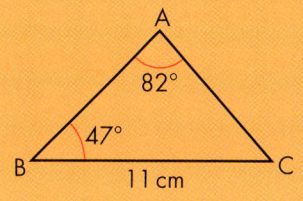

a) Find the lengths of AC and AB.

b) Calculate the area of triangle ABC.

Chief Examiner Says

When you are asked to find lengths or angles in a triangle without a right-angle, you need to use the sine or cosine rule.

Here is an exam question ...

A ship sails from a port P a distance of 7 km on a bearing of 310° and then a further 11 km on a bearing of 070° to arrive at a point X where it anchors.

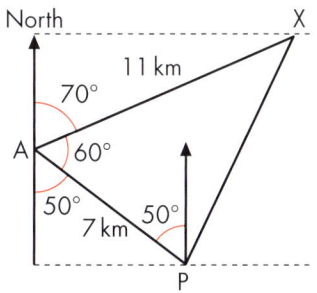

a) Calculate the distance from P to X. **[3]**
b) Calculate how far east of P the point X is. **[2]**

    [3]

... and its solution

a) $PX^2 = 7^2 + 11^2 - 2 \times 7 \times 11 \cos 60°$
   $= 49 + 121 - 77$
   $PX = 9.64 \text{ km}$

b) Distance east $= 11 \sin 70° - 7 \sin 50°$
   $= 4.97 \text{ km}$

Solutions

Test Yourself (1)

a) $\dfrac{a}{\sin A} = \dfrac{b}{\sin B}$

$\dfrac{11}{\sin 82°} = \dfrac{b}{\sin 47°}$

$b = \dfrac{11 \times \sin 47°}{\sin 82°} = 8.12 \text{ cm}$

AC = 8.12 cm

Angle ACB
$= 180° - 47° - 82° = 51°$

$\dfrac{c}{\sin C} = \dfrac{a}{\sin A}$

$c = \dfrac{11 \times \sin 51°}{\sin 82°} = 8.63 \text{ cm}$

AB = 8.63 cm

b) Using $\tfrac{1}{2}ac \sin B$,

Area
$= \tfrac{1}{2} \times 11 \times 8.63 \times \sin 47°$
$= 34.7 \text{ cm}^2$

Now Try This Exam Question

1 X, Y, Z and T are four corners of a cube of side 10 cm.
One corner is sliced off, as shown by the shaded portion, ABC.
XA = 2 cm, BZ = 5 cm, CT = 3 cm

a) Calculate the following lengths.
  i) AB   **ii)** AC   **iii)** BC

b) Calculate the size of angle ACB.   **[10]**

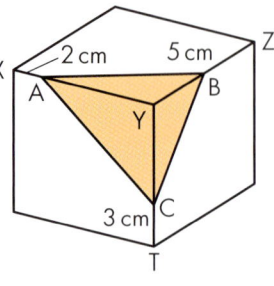

More Exam Practice SE7

Properties of circles

Sectors and segments

- The shaded parts are called the 'minor' sector and the 'minor' segment.
- The unshaded parts are called the 'major' sector and the 'major' segment.

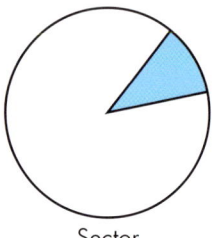

Sector   Segment

Tangent properties of circles

- The tangent at any point on a circle is at right angles to the radius at that point.
- The two tangents to a circle from a point outside a circle are equal in length.

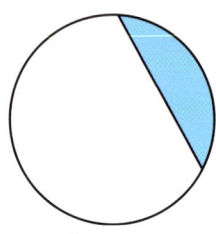

Figure 1

Test Yourself (1)

a) In Figure 1, TA = 9 cm and TO = 10 cm. Find the radius of the circle.

b) Angle ATB = 50°. Find angle TAB.

Chord properties of circles

- The perpendicular from the centre of a circle to a chord, bisects the chord.

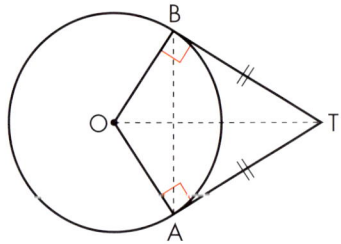

Figure 2

Test Yourself (2)

Angle AOB = 120° and the radius of the circle = 5 cm. Find AB.

More Practice S27

Solutions

Test Yourself (1)

a) Since angle TAO = 90°, use Pythagoras.
$$OA^2 + 9^2 = 10^2$$
$$OA = \sqrt{100 - 81} = 4.36 \text{ cm}$$

b) Since TA = TB, triangle TAB is isosceles.
Angle TAB = $\frac{1}{2}(180 - 50) = 65°$

Test Yourself (2)

Angle BON = $\frac{1}{2}(120°) = 60°$

$\dfrac{BN}{OB}$ = sin 60°, so BN = 5 × sin 60° = 4.33 cm

Since AN = BN, AB = 4.33 × 2 = 8.66 cm

Angle properties of circles

- The angle subtended by an arc at the centre is twice the angle subtended at the circumference.

Figure 3

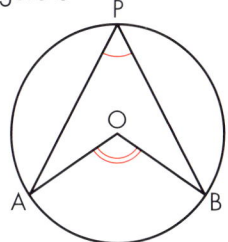

Figure 4

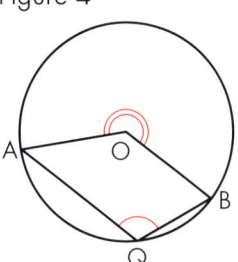

Angle AOB = 2 × angle APB   Reflex angle AOB = 2 × angle AQB

- The angle in a semi-circle is a right-angle.

Figure 5

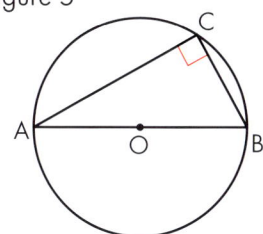

Angle ACB = 90°

- Angles in the same segment are equal.

Figure 6

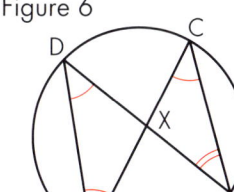

     Angle ADB = angle ACB
and   Angle DAC = angle DBC

- Angles in the alternate segment are equal.

Figure 8

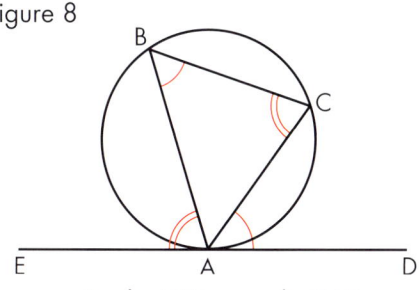

     Angle ABC = angle CAD
and   Angle BAE = angle ACB

Test Yourself (1)

1 In Figure 3, angle AOB = 68°. Find angle APB.

2 In Figure 5, angle CAO = 38° and O is the centre of the circle. Find angle CBA.

3 In Figure 7, angle BCD = 102°. Find angle BAD. Give your reasons.

More Practice S28

- Opposite angles of a cyclic quadrilateral add up to 180°.

Figure 7

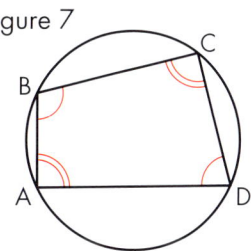

     Angle ABC + angle CDA = 180°
and   Angle BAD + angle BCD = 180°

Chief Examiner Says

When giving reasons, use the standard phrases, e.g.
'angles in a semicircle',
'angle at centre = twice angle at circumference',
'angles in the same segment',
'opposite angles of a cyclic quadrilateral'.

Solutions

Test Yourself (1)

1 Angle APB = $\frac{1}{2}$ angle AOB = $\frac{1}{2}$ × 68 = 34°

2 Angle ACB = 90°
So angle CBA = 180 − 38 − 90 = 52°

3 Angle BAD = 180 − 102 = 78° because angles BCD and BAD are opposite angles of a cyclic quadrilateral.

Here is an exam question ...

O is the centre of the circle and angle ATB is 50°. TA and TB are tangents.

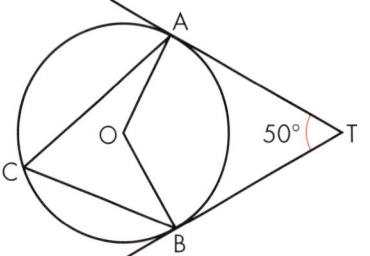

a) Find angle AOB.

b) Find angle ACB, giving a reason for your answer. **[5]**

... and its solution

a) Angle OAT = angle OBT = 90°
Angle AOB = 360 − 90 − 90 − 50 = 130°

b) Angle ACB = $\frac{1}{2}$ × 130 = 65°
Because angle at centre = twice angle at circumference.

Now Try This Exam Question

1 A, B, C and D are points on a circle, centre O.
TA is a tangent to the circle. Find the size of each of the angles labelled x, y and z in the diagram.
Write down the circle property you used to obtain your answer. **[6]**

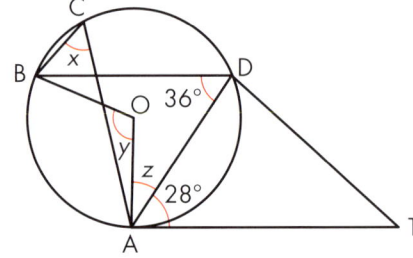

More Exam Practice SE8

Three-dimensional shapes

Pyramids, cones and spheres

- For a pyramid, the shape of the base is usually part of its name.
  For example 'a square-based pyramid'.
  Volume, $V = \frac{1}{3}$ × area of base × height

- A 'pyramid' with a circular base is a cone
  Volume, $V = \frac{1}{3}\pi r^2 h$.
  Curved surface area, $A = \pi r l$

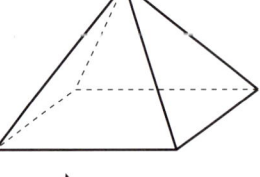

- For a sphere:
  $V = \frac{4}{3}\pi r^3$
  Surface area = $4\pi r^2$

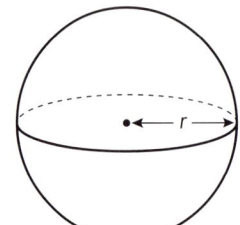

Test Yourself (1)

1 A pyramid has a square base of side 4 m and is 9 m high. Calculate the volume of the pyramid.

2 Find the height of a cone with volume 2.5 litres and base radius 10 cm.

3 A sphere has a surface area of 500 cm². Calculate the radius of the sphere.

More Practice S29

Solutions

Test Yourself (1)

1 $V = \frac{1}{3} \times (4 \times 4) \times 9$
$= 48 \text{ m}^3$

2 $2500 = \frac{1}{3} \times \pi \times 10^2 \times h$
$h = \frac{3 \times 2500}{\pi \times 100}$
$h = 23.9 \text{ cm}$

3 $500 = 4\pi r^2$
$r^2 = \frac{500}{4\pi}$
$r = 6.3 \text{ cm}$

Chief Examiner Says

Don't forget that $4\pi r^2$ is 4π times r^2, not $(4\pi r)^2$.

Volume and surface area of similar shapes

For similar shapes:
- Area scale factor = (length scale factor)^2.
- Volume scale factor = (length scale factor)^3.

Test Yourself (1)

Two similar cylinders have heights 8 cm and 16 cm.

a) The smaller cylinder has a volume of 60 cm^3. Find the volume of the larger cylinder.

b) Another similar cylinder has a volume of 202.5 cm^3. Find its height.

More Practice S30

Here is an exam question ...   ... and its solution

A hemispherical bowl has a radius of 30 cm.

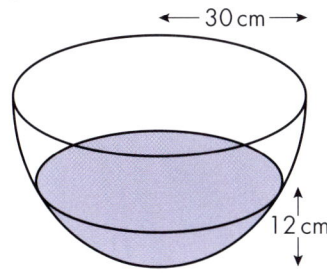

←— 30 cm —→

12 cm

a) Calculate the volume of the bowl. Leave your answer as a multiple of π.

b) Water is poured into the bowl to a depth of 12 cm. Calculate the radius of the surface of the water.   **[6]**

a) $v = \frac{1}{2} \times \frac{4}{3} \times \pi \times 30^3$

$= \frac{1}{2} \times \frac{4}{3} \times \pi \times 27\,000$

$= 2 \times \pi \times 9000$

$= 18\,000\pi \text{ cm}^3$

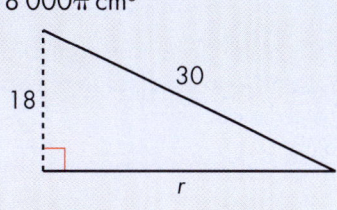

b) $r^2 = 30^2 - 18^2$

$= 576$

$r = 24 \text{ cm}$

Now Try These Exam Questions

1 An ice-cream cone has height 11 cm and radius 3.5 cm. Ice-cream completely fills the cone and forms a hemisphere on the top of the cone. Neglecting the thickness of the cone, calculate the volume of ice-cream.

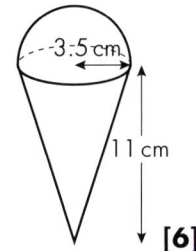

--3.5 cm--

11 cm

[6]

2 Three similar suitcases each have dimensions which are 25% greater than those of the next smaller suitcase.

a) The middle suitcase is 40 cm tall. How tall is the smallest suitcase?

b) The middle suitcase has a volume of 32 000 cm^3. Calculate the volume of the largest suitcase.   **[6]**

More Exam Practice SE9

Solutions

Test Yourself (1)

a) Length scale factor $= \frac{16}{8} = 2$

Volume $= 60 \times 2^3 = 480 \text{ cm}^3$

b) Length scale factor $= \left(\frac{202.5}{60}\right)^{\frac{1}{3}} = 1.5$

Height $= 8 \times 1.5 = 12 \text{ cm}$

Transformations, coordinates and vectors

Enlargement

● If the scale factor is negative, each point is transformed to a point on the other side of the centre of enlargement.

Test Yourself (1)

Draw axes from -12 to 6 for x and -8 to 6 for y. Plot the points (2, 3), (4, 3) and (2, 6) and join them to form a triangle. Enlarge the triangle by scale factor -2 with centre $(-1, 2)$.

More Practice S31

Finding the single transformation equivalent to two given transformations

● Perform each of the given transformations in sequence.
Check the object and the final image to determine the equivalent single transformation.

Test Yourself (2)

Draw axes from -8 to 6 for x and -8 to 4 for y. Plot the points (4, 2), (4, 3), (4, 4) and (5, 3) and join them to form a flag. Reflect the flag in the line $x = 1$ and reflect the image in the line $y = -2$. Find the single transformation that is equivalent to these two reflections.

Chief Examiner Says

Don't forget to give all the information needed to define the transformation.

Solutions

Test Yourself (1)

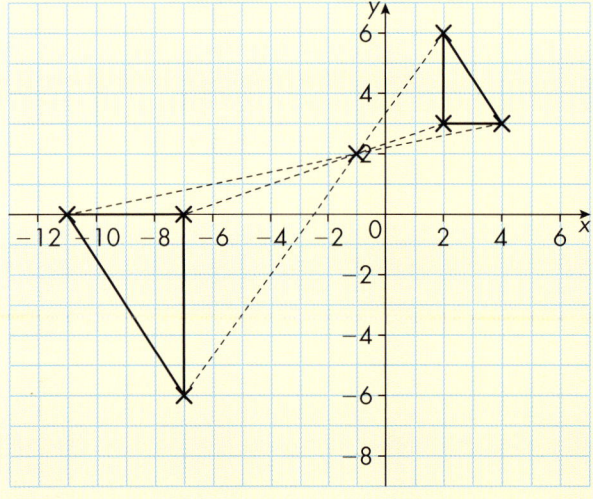

Image has coordinates $(-7, 0)$, $(-11, 0)$ and $(-7, -6)$.

Test Yourself (2)

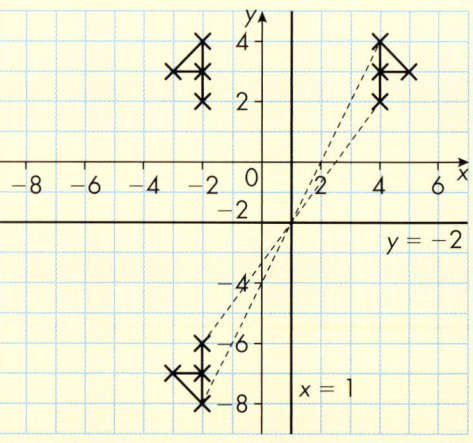

Final image has coordinates $(-2, -6)$, $(-2, -7)$, $(-2, -8)$ and $(-3, -7)$. The transformation is a rotation of 180° about a centre $(1, -2)$.

Calculating with coordinates

- To find the length of the line joining two points, use Pythagoras' theorem, subtracting the coordinates to find the two shorter sides.

More Practice S32

Test Yourself (1)

When A is (1, 5) and B is (7, 2), find the length AB.

Length, area and volume

- To distinguish between formulae for length, area and volume, consider the dimensions of the formula.

Chief Examiner Says

π is a number, not a length.

Test Yourself (2)

In these formulae b, h, x, y and r represent lengths.
a) Which of these formulae represent a length?
   i) $\frac{1}{2}bh$ **ii)** $3b$ **iii)** $b + 2h$
b) Which of these formulae represent an area?
   i) xy **ii)** xy^2 **iii)** $x(x + y)$
c) Which of these formulae represent a volume?
   i) r^3 **ii)** $\pi r^2 h$ **iii)** $r^2(r + h)$

More Practice S33

Vector geometry

- Vectors are added by starting the second vector where the first one finishes.
- To subtract two vectors, use $\mathbf{a} - \mathbf{b} = \mathbf{a} + -\mathbf{b}$.
- The resultant of two vectors $\mathbf{a}$ and $\mathbf{b}$ is $\mathbf{a} + \mathbf{b}$.
- If they are in component form, vectors may be added by adding their corresponding components.
- $\mathbf{a} = k\mathbf{b}$ means that $\mathbf{a}$ is parallel to $\mathbf{b}$ and it is k times as long.
- $\overrightarrow{AB} = k\overrightarrow{AC}$ means that A, B and C are in a straight line and the length of AB = k times the length of AC.

More Practice S34

Test Yourself (3)

If $\mathbf{a} = \begin{pmatrix} 3 \\ 4 \end{pmatrix}$, $\mathbf{b} = \begin{pmatrix} 2 \\ -1 \end{pmatrix}$, find $\mathbf{a} + \mathbf{b}$.

Solutions

Test Yourself (1)

$AB^2 = 6^2 + 3^2 = 45$
$AB = \sqrt{45} = 6.7$ units to 1 d.p.

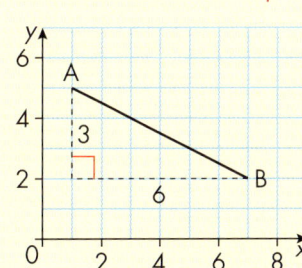

Test Yourself (2)

a) ii) and **iii)**
b) i) and **iii)**
c) i), **ii)** and **iii)**

Test Yourself (3)

$\mathbf{a} + \mathbf{b} - \begin{pmatrix} 3 + 2 \\ 4 - 1 \end{pmatrix} - \begin{pmatrix} 5 \\ 3 \end{pmatrix}$

Here is an exam question ...

Shape A is translated by $\begin{pmatrix} 2 \\ -2 \end{pmatrix}$. Its image is B.

Shape B is enlarged with centre (1, 0) and scale factor −3.
Its image is C.
Describe the transformation which maps C onto A.

[7]

... and its solution

The transformation which maps C onto A is an enlargement with centre (−0.5, 1.5) and scale factor $-\frac{1}{3}$.

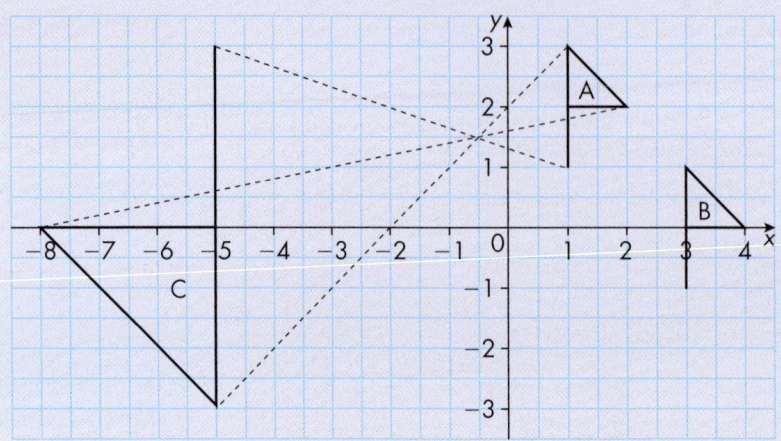

Now Try These Exam Questions

1 Describe the enlargement which is equivalent to a rotation through 180° about (0, 0). **[3]**

2 A and C have coordinates (−2, 1) and (2, −2) respectively. $\overrightarrow{AB} = \begin{pmatrix} 6 \\ 5 \end{pmatrix}$. M is the midpoint of $\overrightarrow{AB}$ and N is the midpoint of $\overrightarrow{AC}$.

a) Write down the coordinates of B, M and N.

b) Calculate the vectors $\overrightarrow{BC}$ and $\overrightarrow{MN}$.

c) State the relationship between BC and MN. **[8]**

More Exam Practice SE10

Measures

Upper and lower bounds of combined measurements

- To find the upper bound of a sum, add the two upper bounds.
- To find the lower bound of a sum, add the two lower bounds.
- To find the upper bound of a difference, subtract the lower bound from the upper bound.
- To find the lower bound of a difference, subtract the upper bound from the lower bound.
- To find the upper bound of a product, multiply the upper bounds.
- To find the lower bound of a product, multiply the lower bounds.
- To find the upper bound of a division, divide the upper bound by the lower bound.
- To find the lower bound of a division, divide the lower bound by the upper bound.

Test Yourself (1)

a) A box of apples weighs 5 kg to the nearest kg.
What are the upper and lower bounds of the weight of 10 of these boxes?

b) A rectangle has sides 3.5 cm and 4.6 cm measured to 2 s.f. Find the minimum and maximum value of the following:
   i) The perimeter
   ii) The area.

c) $A = \dfrac{3b}{c}$, $b = 3.62$, $c = 5.41$ to 2 d.p.
Find the lower and upper bounds of A.

More Practice S35

Here is an exam question ...   ... and its solution

When a ball is thrown upwards, the maximum height, h, it reaches is given by $h = \dfrac{U^2}{2g}$.

It is given that $U = 4.2$ and $g = 9.8$, both correct to 2 s.f. Calculate the upper and lower bounds of h. **[6]**

To find the upper bound of h, use the upper bound of U and the lower bound of g.
To find the lower bound of h use the opposite.

Upper $h = \dfrac{4.25^2}{2 \times 9.75}$
$= 0.926\,282$
$= 0.93$ to 2 s.f.

Lower $h = \dfrac{4.15^2}{2 \times 9.85}$
$= 0.874\,238\,5$
$= 0.87$ to 2 s.f.

Now Try This Exam Question

1 A formula used in science is $a = \dfrac{v - u}{t}$.

$u = 17.4$, $v = 30.3$ and $t = 2.6$, all measured correct to the nearest 0.1.
Find the maximum possible value of a. **[4]**

More Exam Practice SE11

Solutions

Test Yourself (1)

a) Upper bound $= 10 \times 5.5 = 55$ kg
Lower bound $= 10 \times 4.5 = 45$ kg

b) i) Minimum perimeter $= 2 \times 3.45 + 2 \times 4.55 = 16$ cm
Maximum perimeter $= 2 \times 3.55 + 2 \times 4.65 = 16.4$ cm

   ii) Minimum area $= 3.45 \times 4.55 = 15.6975 = 15.7$ cm^2
Maximum area $= 3.55 \times 4.65 = 16.5075 = 16.5$ cm^2

c) Lower bound of $A = \dfrac{3 \times 3.615}{5.415} = 2.0028 = 2.00$

Upper bound of $A = \dfrac{3 \times 3.625}{5.405} = 2.0120 = 2.01$

Chief Examiner Says

Remember, the value must be less than the upper bound. However, do not write something smaller, e.g. 5.4999.

Arcs and sectors of circles

Arc length and area of a sector of a circle

- Arc length $= \dfrac{x}{360} \times 2\pi r$

- Area of sector $= \dfrac{x}{360} \times \pi r^2$

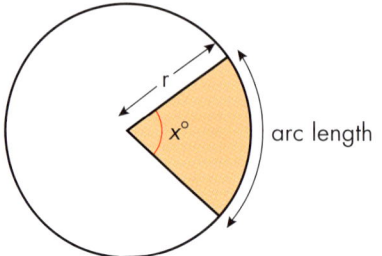

Chief Examiner Says

Sector area is directly proportional to the angle at the centre.

More Practice S36

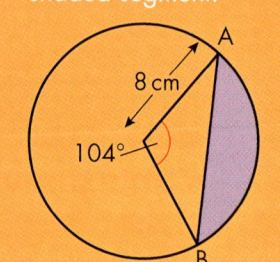

Test Yourself (1)

a) Find the arc length AB.

b) Find the area of the shaded segment.

Here is an exam question ...

A piece of card is cut from a circle of radius 25 cm as shown. The remaining card is folded so that the straight edges meet to make a hat with a circular base.

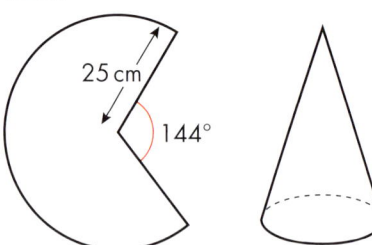

a) What is the surface area of the hat?

b) What is the circumference of the base of the hat? **[6]**

... and its solution

a) Surface area $= \dfrac{216}{360} \times \pi \times 25^2$

$= 1178.1 \text{ cm}^2$

b) Circumference $= \dfrac{216}{360} \times 2 \times \pi \times 25$

$= 94.2 \text{ cm}$

Now Try This Exam Question

1 The diagram shows a sector of a circle. The angle is 150° and the arc length is 20 cm. Find the radius of the circle.

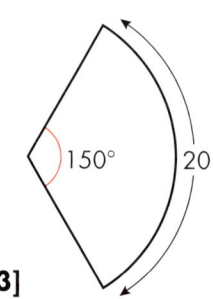

[3]

More Exam Practice SE12

Solutions

Test Yourself (1)

a) Arc length $= \dfrac{104}{360} \times \pi \times 16 = 14.5 \text{ cm}$

b) Segment $=$ sector $-$ triangle

$= \dfrac{104}{360} \times \pi \times 8^2 - \dfrac{1}{2} \times 8 \times 8 \times \sin 104°$

$= 27.0 \text{ cm}^2$

Processing and representing data

Histograms

- Histograms are particularly useful when you are dealing with unequal intervals.
- A histogram is similar to a frequency diagram except that on a histogram the area of the bar is equal to the frequency.
  - Width of interval $\times$ height of bar = frequency
  - Height = $\dfrac{\text{frequency}}{\text{width of interval}}$
- The height is called the frequency density.

More Practice HD11

Test Yourself (1)

The table shows the distribution of times (t seconds) that a company took to answer 100 telephone calls.

Time (t seconds)	Frequency (f)	Frequency density = $f \div$ width
$0 < t \leqslant 2$	28	
$2 < t \leqslant 5$	48	
$5 < t \leqslant 10$	12	
$10 < t \leqslant 15$	7	
$15 < t \leqslant 20$	5	

a) Complete the column for frequency density.

b) Draw a histogram to represent the data.

Time series and moving averages

- A time series is simply a graph where the x-axis is time and the y-axis is any quantity which varies with time. For example, quarterly sales of a company.
- A moving average is used where the graph shows a cyclical change. For example, higher sales always in the Summer, lower sales always in the Winter.
- To calculate the moving average in the example of quarterly sales:
  1 Find the mean for the four quarters in the first year.
  2 Omit the first quarter of year one and include the first quarter of year two and find the new mean.
  3 Omit the second quarter of year one and include the second quarter of year two and so on.
- The moving average is plotted in the middle of the four quarters.

More Practice HD12

Test Yourself (2)

The table gives the sales figures in £millions for a company.

Quarter	1st	2nd	3rd	4th
1998	4.2	5.6	7.2	4.9
1999	3.9	5.5	6.8	4.7
2000	3.8	5.2	6.5	4.6

a) Plot a time series graph.

b) Calculate the four-quarter moving averages and plot them on the graph.

Solutions

Test Yourself (1)

a) Frequency densities: 14, 16, 2.4, 1.4, 1.

b)

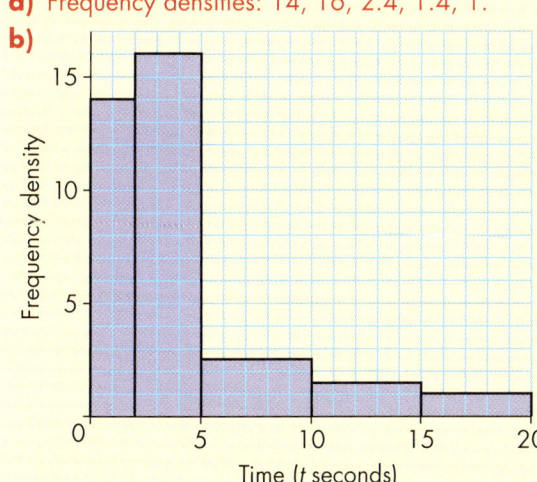

Test Yourself (2)

a)

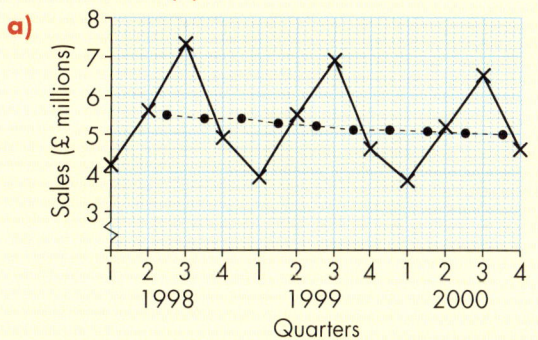

b) $(4.2 + 5.6 + 7.2 + 4.9) \div 4 = 5.475$
$(5.6 + 7.2 + 4.9 + 3.9) \div 4 = 5.4$
$(7.2 + 4.9 + 3.9 + 5.5) \div 4 = 5.375$
$(4.9 + 3.9 + 5.5 + 6.8) \div 4 = 5.275$
$(3.9 + 5.5 + 6.8 + 4.7) \div 4 = 5.225$
$(5.5 + 6.8 + 4.7 + 3.8) \div 4 = 5.2$
$(6.8 + 4.7 + 3.8 + 5.2) \div 4 = 5.125$
$(4.7 + 3.8 + 5.2 + 6.5) \div 4 = 5.05$
$(3.8 + 5.2 + 6.5 + 4.6) \div 4 = 5.025$

Cumulative frequency graphs, median, quartiles, interquartile range and box plots

- Cumulative frequency is the running total of the frequencies in a distribution. The last cumulative frequency is the total of the frequencies and is often given in the question.
- Cumulative frequency is plotted at the upper bound of each interval. The points can be joined by a curve or by straight lines. The graph shows a fairly typical shape for a cumulative frequency curve.
- To find the median of a frequency distribution, draw a line across the graph at half the total frequency to meet the curve and then down to read off the value on the horizontal-axis.
- To find the quartiles of a frequency distribution, draw lines across the graph at a quarter and three-quarters of the total frequency to meet the curve and then down to read off the values on the horizontal-axis.
- Interquartile range (upper quartile − lower quartile) is a measure of spread.
- Box plots are a useful way of displaying data. They are also called 'box and whisker' plots.

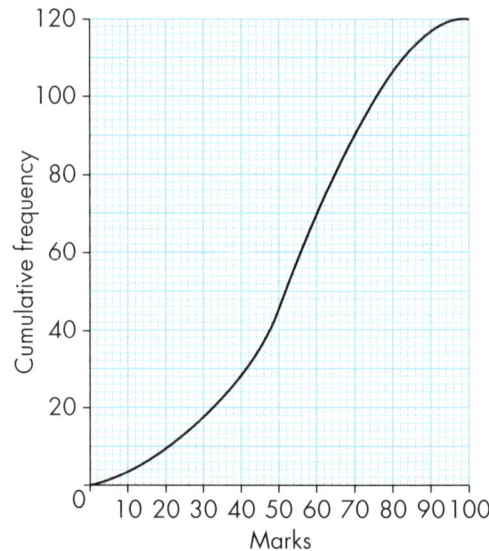

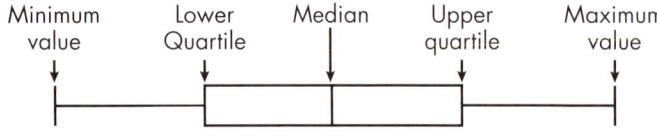

Test Yourself (1)

The table shows the distribution of the masses (m g) of 80 tomatoes.

a) Complete the right-hand column to shows the cumulative frequencies.

b) Draw a cumulative frequency diagram.

c) Estimate the number of tomatoes weighing more than 93 g.

d) Find the median and interquartile range.

e) Draw a box-and-whisker plot.

Mass (g)	Frequency	Cumulative frequency
$70 < m \leqslant 75$	4	
$75 < m \leqslant 80$	10	
$80 < m \leqslant 85$	22	
$85 < m \leqslant 90$	27	
$90 < m \leqslant 95$	14	
$95 < m \leqslant 100$	3	

More Practice HD13

Solutions

Test Yourself (1)

a)

Mass (g)	Frequency	Cumulative frequency
$70 < m \leqslant 75$	4	4
$75 < m \leqslant 80$	10	14
$80 < m \leqslant 85$	22	36
$85 < m \leqslant 90$	27	63
$90 < m \leqslant 95$	14	77
$95 < m \leqslant 100$	3	80

b)

c) $80 - 71 = 9$

d) Median = 86 g

Interquartile range = $89.5 - 81.5 = 8$ g

e)

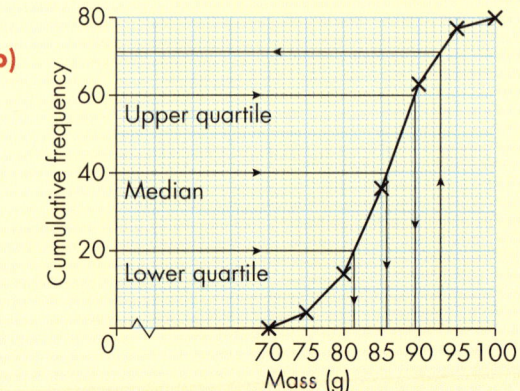

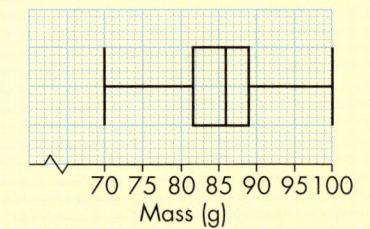

Comparing distributions

- If you are asked to compare two distributions, make two basic comparisons:
  - For the first comparison, decide which distribution is bigger on average. Evidence of this is a higher mean or a higher median or a higher mode. In a frequency diagram or polygon, a higher mode is indicated by the highest bar or highest point being further to the right.
  - For the second comparison, decide which distribution has the greater spread. This is indicated by a bigger range or interquartile range. If you know the interquartile ranges, it is better to compare those.

Test Yourself (1)

This box plot summaries the times, in minutes, taken for 100 students at Fairlands College to get to school.

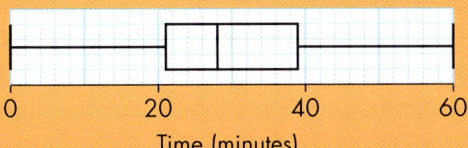

Time (minutes)

The table below summarises the times, in minutes, taken for the students at Breidon school to get to school.

Minimum	Lower quartile	Median	Upper quartile	Maximum
5	15	25	41	80

a) On the same scale as above, draw a box plot for Breidon School.

b) Make two comparisons between the distribution of times for the two schools.

More Practice HD14

Solutions

Test Yourself (1)

a)

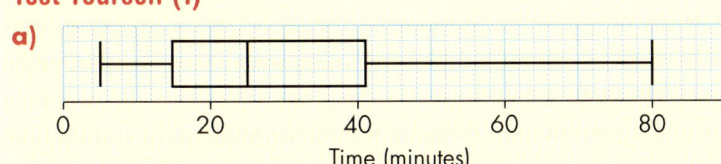

Time (minutes)

b) Two valid comparisons such as:
1. The median shows that the Fairlands took longer on average to get to school.
2. The times are more consistent at Fairlands as the interquartile range was only 18, whereas at Breidon it was 26 (or use the ranges 60 and 75 to draw the same conclusion).

Here is an exam question ...   ... and its solution

A test was carried out to establish the ability of a mouse to find food. The test was carried out on 120 mice. The distribution of times taken to reach the food is given in the table.

Time (t seconds)	Frequency
0 < t ≤ 10	18
10 < t ≤ 15	46
15 < t ≤ 20	35
20 < t ≤ 30	13
30 < t ≤ 50	8

Draw a histogram to represent this information.   **[4]**

Time (t seconds)	Frequency	Frequency density
0 < t ≤ 10	18	1.8
10 < t ≤ 15	46	9.2
15 < t ≤ 20	35	7
20 < t ≤ 30	13	1.3
30 < t ≤ 50	8	0.4

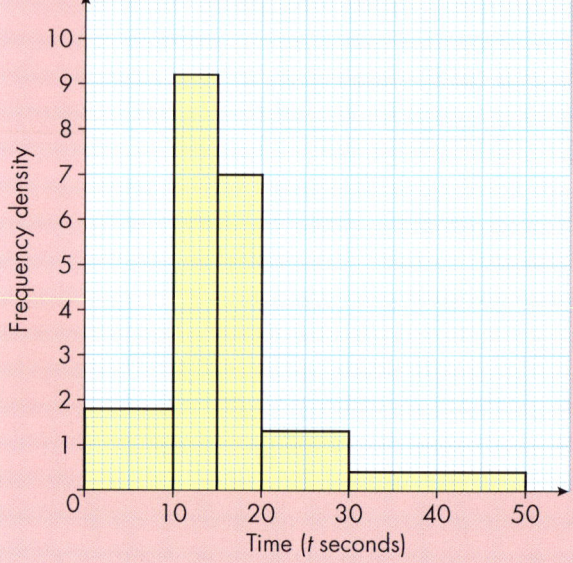

Now Try These Exam Questions

1   The weight loss in pounds of each of 80 members of a slimming club is summarised in the table.

Weight loss (w pounds)	Frequency
0 < w ≤ 5	9
5 < w ≤ 10	13
10 < w ≤ 15	21
15 < w ≤ 20	17
20 < w ≤ 25	10
25 < w ≤ 30	8
30 < w ≤ 35	2

   a) i) Copy and complete the cumulative frequency table.   **[1]**

Weight loss (w pounds)	Cumulative frequency
w ≤ 5	9
w ≤ 10	
w ≤ 15	
w ≤ 20	
w ≤ 25	
w ≤ 30	
w ≤ 35	

   ii) Draw the cumulative frequency graph for the weight loss of the club. Use a scale of 2 cm to 5 pounds on the horizontal axis and 1 cm to 10 members on the vertical axis.   **[3]**

   b) Use your graph to find the following.
   i)   The median weight   **[1]**
   ii) The interquartile range   **[2]**

   c) The company running the club guarantees that everyone will lose at least 8 pounds. Use your graph to estimate how many members achieve this target.   **[2]**

   d) The club's slimming champion lost 30% of his weight and now weighs 84 kg. Calculate the champion's weight before he started slimming.   **[3]**

   e) Draw a box plot to summarise the distribution.   **[3]**

Now Try These Exam Questions (contd.)

2 A doctor's patients are divided by ages as shown in the table.

Age (*x*) in years	Number of calls
$0 < x \leqslant 5$	14
$5 < x \leqslant 15$	41
$15 < x \leqslant 25$	59
$25 < x \leqslant 45$	70
$45 < x \leqslant 75$	16

Draw a histogram to represent this information. Use a scale of 2 cm to 10 years on the horizontal-axis and an area of 1 cm² to represent five patients. **[3]**

3 The table shows a company's quarterly sales of umbrellas in the years 2003 to 2006. The figures are in thousands of pounds.

	1st quarter	2nd quarter	3rd quarter	4th quarter
2003	153	120	62	133
2004	131	105	71	107
2005	114	110	57	96
2006	109	92	46	81

a) Plot these figures in a graph. Use a scale of 1 cm to each quarter on the horizontal axis and 2 cm to 20 thousand pounds on the vertical axis. **[3]**

b) The four quarter moving averages are *a*, *b*, 107.5, 110, 104.5, 99.25, 100.5, 97, 94.25, 93, 88.5, 85.75 and 82.

Calculate the values of *a* and *b*, the first two moving averages. **[2]**

c) Plot all the moving averages on your graph. **[2]**

d) Comment on the general trend and the quarterly variation. **[2]**

e) Draw a trend line for the points representing the moving averages. **[1]**

f) Use your trend line to predict the next moving average (2nd quarter of 2006 to the 1st quarter of 2007). **[1]**

g) Use the value you found in part (f) to predict the umbrella sales for the first quarter of 2007. **[2]**

4 A hospital recorded the birth weights, in kilograms, of 100 boys and 100 girls. The weights are summarised in the table.

a) Draw two box-and-whisker diagrams to show this information. **[6]**

b) Use your diagrams to compare the birth weights of boys and girls. **[2]**

	Girls	Boys
Median	3.2	3.3
Lower quartile	2.6	2.3
Upper quartile	3.8	4.0
Minimum	1.2	1.2
Maximum	4.4	4.7

More Exam Practice HDE5

Probability

Combining probabilities

- Equally likely outcomes may be listed in a table or shown on a grid
- For mutually exclusive events P(A or B) = P(A) + P(B).
- For independent events P(A and B) = P(A) × P(B).
- When outcomes are not equally likely, use **tree diagrams**. Remember:
    - each set of branches shows the possible outcomes of the event.
    - the probabilities on each set of branches should add to 1.
    - when events are **independent**, the outcomes of the second event are not affected by the outcomes of the first.
    - to find the probabilities of the combined events, multiply the probabilities along the branches.

Chief Examiner Says

- If you are finding the probability of all but one of the outcomes, it is often easier to work out 1 − P(remaining outcome).

Chief Examiner Says

- When listing possibilities, be systematic to make sure you don't miss any possibilities.

Chief Examiner Says

- One of the most common errors is to add probabilities instead of multiplying them. If you see the word 'and' or 'both' or 'all' then multiply the probabilities.

Test Yourself (1)

This tree diagram shows the probabilities that Penny has to stop at lights or a level crossing on her way to work.

a) Complete the tree diagram.
b) What is the probability that
   i) she does not stop at either the lights or the level crossing?
   ii) she stops at either the lights or the level crossing but not both?

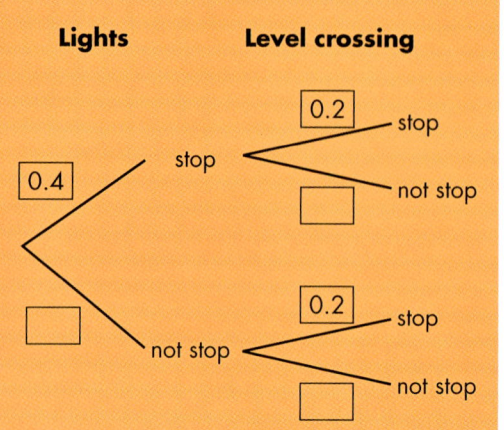

More Practice HD15

Conditional probabilities

- When events are not independent, the outcome of one affects the probability that the other happens.
- In a tree diagram in this case, the probabilities on the second pairs of branches will be different.

More Practice HD16

Test Yourself (2)

The probability that Pali wakes up late on a work morning is 0.1. When he wakes up late, the probability that he misses the bus is 0.8. When he doesn't wake up late, the probability that he misses the bus is 0.2. Draw a tree diagram to represent this, and find the probability that he misses the bus on a work morning.

Solutions

Test Yourself (1)

Chief Examiner Says

Even if it is not asked for, it is a good idea to draw a tree diagram.

Test Yourself (2)

Chief Examiner Says

Remember that, at each stage, the sum of the probabilities is 1.

a)

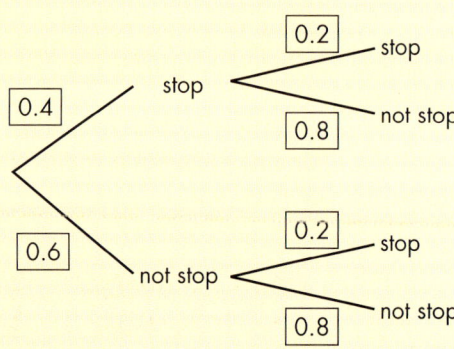

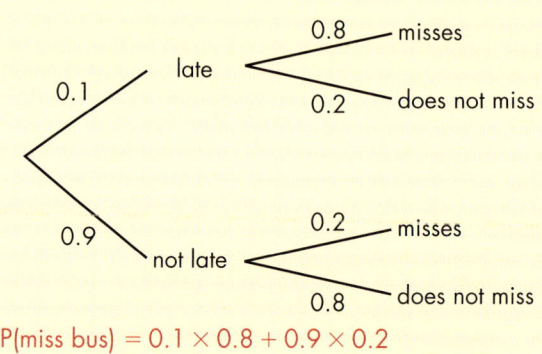

$$P(\text{miss bus}) = 0.1 \times 0.8 + 0.9 \times 0.2$$
$$= 0.08 + 0.18 = 0.26$$

b) i) P(does not stop at either) = $0.6 \times 0.8 = 0.48$
   ii) P(stops at either but not both) = $0.4 \times 0.8 + 0.6 \times 0.2 = 0.32 + 0.12 = 0.44$

Here is an exam question ...

Pete likes crisps. Without looking, he picks a bag out of an assorted pack. There are 12 bags of crisps in the pack. Two of these bags are ready-salted. The manufacturers say that one bag in every 100 has a gold reward in it. This is independent of the flavour of the crisps.

a) Draw a tree diagram to show the probabilities. **[3]**

b) What is the probability that Pete picks a ready-salted bag with a gold reward in it? **[2]**

c) What is the probability that Pete picks either a ready-salted bag or one with a gold reward in it but not both of these? **[3]**

... and its solution

a)

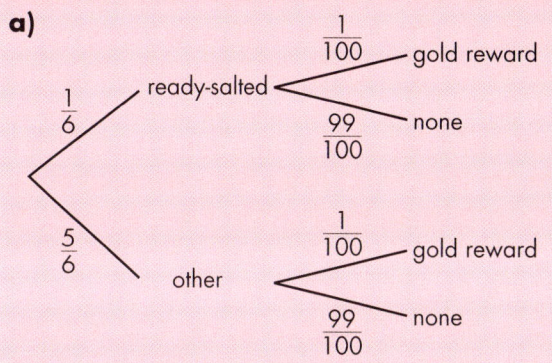

b) P(picks ready-salted with gold reward)
$= \frac{1}{6} \times \frac{1}{100} = \frac{1}{600}$

c) P(picks ready-salted or gold reward but not both) $= \frac{1}{6} \times \frac{99}{100} + \frac{5}{6} \times \frac{1}{100} = \frac{104}{600} = \frac{13}{75}$

Now Try These Exam Questions

1 Ellen and Schweta are practising goal scoring in netball. When they each take one shot the probability that Ellen scores a goal is 0.7 and the probability that Schweta scores a goal is 0.8. These probabilities are independent.

a) Complete the tree diagram. **[2]**

Ellen   Schweta

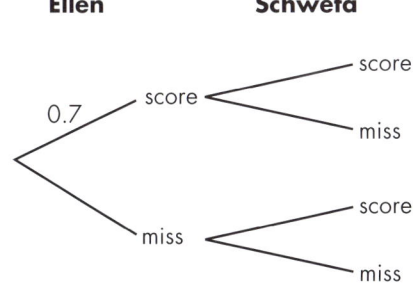

b) Calculate the probability that, at their next attempt:
  i) both Ellen and Schweta score a goal. **[2]**
  ii) either Ellen or Schweta, but not both, scores a goal. **[3]**

2 Box A contains pens. 20% are red, 50% are blue and 30% are black. Box B contains pencils. 50% are red, 15% are blue and 35% are black. Jack takes a pen from Box A and a pencil from Box B at random.

a) Draw a tree diagram to show the possible outcomes. **[3]**

b) Work out the probability that he takes a red pen and a red pencil. **[2]**

c) Work out the probability that he takes a pen and pencil of the same colour. **[3]**

3 In this question express your answers as fractions in their lowest terms.
There are 30 students in a class.
The table gives information about boys and girls and whether they are left-handed or right-handed.

	Right-handed	Left-handed	Total
Boys	15	3	18
Girls	10	2	12
Total	25	5	30

a) A student is chosen at random. Given that the student is a girl, what is the probability that she is left-handed? **[2]**

b) Two students are chosen at random. What is the probability that they are left-handed? **[2]**

c) Three of the girls are chosen at random. What is the probability that exactly two of them are right handed? **[3]**

More Exam Practice HDE6

Sampling

- To investigate something about a population you may use a census or a sample.
- In a census you find out the information about every member of the population.
- In a sample the information is obtained from a small proportion of the population, usually 10% or less. You need to ensure that the sample is representative of the whole population.
- There are various methods of sampling, but the two here are all you need to know.
- In either case you need to decide on the size of sample you require. This is usually a percentage, for example 10%, but it could be a fixed number.
- The advantage of a sample is that it is cheaper and quicker.
- The disadvantage is that, since it is not a census, its accuracy depends on how representative the sample is.

Simple random sampling

- For simple random sampling, every member of the population must have an equal chance of being chosen, irrespective of who has already been chosen.
- Give every member of a population a number and choose numbers at random.
- Random numbers can be chosen by a physical method, for example putting the numbers in a bag and choosing at random, or by using a random number generator on a calculator or computer.

Stratified sampling

- In representative stratified sampling, you set up strata that you want to ensure are fairly represented in your sample.
- For example, in a school you may want to ensure that every year group is represented fairly in the sample. So your strata will be the year groups.
- You then choose within those year groups using simple random sampling. So you may, for instance, choose at random 10% of each year group.

Test Yourself (1)

Adam is carrying out a survey about use of public transport in his area. He decides to survey the first person on every page of the phone book.

Why is this **not** a simple random sample?

More Practice HD17

Solutions

Test Yourself (1)

Everyone does not have an equal chance of being chosen. Some people are ex-directory, etc.

Chief Examiner Says

The questions on this topic almost always ask you to describe a method or make comments. It is worth making sure you know the statements in the above bullet points.

Here is an exam question ...

A homework survey is to be carried out in a large school of 1700 students. Because of the school's size, it is impractical to obtain the view of each student. You decide to sample the students to obtain an accurate picture of times spent on homework.

List three steps you would take to ensure that you would obtain a representative sample. **[3]**

... and its solution

1 Make sure sample is big enough (e.g. 10%)
2 Make sure all ages/year groups are represented (e.g. stratified)
3 Make sure both genders are represented

Or some description of random sampling.

Chief Examiner Says

Make sure you read the question carefully. This question is about the sample, not about the questionnaire, conditions of questioning, etc.

Now Try These Exam Questions

1 Fiona, Raiza and Simon conduct a survey on the way students travel to their school.
To do this they each decide to take a 10% sample.
The school has 800 students.

   a) Give one advantage and one disadvantage of using a sample to obtain the data. **[2]**

   b) Fiona decides to go outside the school gate at 8.45 am and ask the first 80 students who arrive.
   Give one reason why this is not a good way to obtain the sample. **[1]**

   c) Raiza decides to take a simple random sample of 10% of the students.
   Describe a way in which Raiza might select her random sample. **[2]**

   d) Simon decides to take a stratified random sample.
   Give one advantage this may have over a simple random sample and suggest possible strata. **[2]**

2 You are doing a survey about kitchen appliances. You are going to sample the households in a small town. You decide to take a stratified sample of 10% of the households.

   a) Describe carefully how you would choose your stratified sample. **[2]**

   b) State one advantage your method has over a simple random sample **[1]**

   c) State one advantage your method has over taking every 10th name from the local telephone directory. **[1]**

More Exam Practice HDE7

Solutions to Now Try These Exam Questions

Section 1

Number: Integers (page 2)

1 a) 64 **b)** 125 **c)** 256

2 a) $30 = 2 \times 15$
  $= 2 \times 3 \times 5$

  b) 3

3 $10 = 2 \times 5$
  $12 = 2 \times 2 \times 3$
  $20 = 2 \times 2 \times 5$
  HCF $= 2$
  LCM $= 2 \times 2 \times 3 \times 5$
    $= 60$

Number: Fractions (page 4)

1 a) $\frac{2}{3} + \frac{4}{5} = \frac{10}{15} + \frac{12}{15}$
    $= \frac{22}{15}$
    $= 1\frac{7}{15}$

  b) $\frac{\cancel{12}}{\cancel{3}} \times \frac{\cancel{6}}{\cancel{12}_2} = \frac{1}{1} \times \frac{1}{2} = \frac{1}{2}$

2 a) $\frac{3}{4} = \frac{30}{40}, \frac{7}{10} = \frac{28}{40}, \frac{3}{5} = \frac{24}{40}, \frac{5}{8} = \frac{25}{40}$

  Order is $\frac{24}{40}, \frac{25}{40}, \frac{28}{40}, \frac{30}{40}$ that is $\frac{3}{5}, \frac{5}{8}, \frac{7}{10}, \frac{3}{4}$

  b) $\frac{24}{40} + \frac{25}{40} + \frac{28}{40} + \frac{30}{40} = \frac{107}{40} = 2\frac{27}{40}$

3 a) $2\frac{3}{8} - 1\frac{1}{2} = 1\frac{3}{8} - \frac{4}{8}$
    $= \frac{8}{8} + \frac{3}{8} - \frac{4}{8}$
    $= \frac{7}{8}$

  b) $\frac{2}{3} \div \frac{4}{5} = \frac{2}{3} \times \frac{5}{4}$
    $= \frac{1}{3} \times \frac{5}{2} = \frac{5}{6}$

4 $2\frac{1}{4} - \frac{7}{16} = 2 + \frac{4}{16} - \frac{7}{16}$
  $= 1 + \frac{16}{16} + \frac{4}{16} - \frac{7}{16}$
  $= 1\frac{13}{16}$ inch left

Number: Percentages (page 6)

1 Loss $= 5595 - 4795$
   $= £800$
  Fraction loss $= \frac{800}{5595}$
      $= 0.142\,984\ldots$
  Percentage loss $= 14.30\%$

2 Selling price including VAT $= 6.95 \times 1.175$
          $= 8.166\,25$
          $= £8.17$

3 Fraction occupied $= \frac{271}{320}$
        $= 0.846\,875$
  Percentage occupied $= 84.69\%$

4 Sale price $= 0.85 \times 80$
       $= £68$

5 Value after 6 months $= 185\,000 \times (1.01)^6$
           $= 196\,381.227\,9$
           $= £196\,381$

Number: Ratio (page 9)

1 a) Gold $= 15 \times \frac{7}{3}$
     $= 35\,g$

  b) Total parts $= 10$
    Multiplier $= 20 \div 10$
        $= 2$
    Gold $= 7 \times 2$
      $= 14\,g$

2 Mixed fruit $= 100 \times \frac{25}{10}$
        $= 250\,g$
  Flour $= 250 \times \frac{25}{10}$
     $= 625\,g$

3 Multiplier $= 420 \div 7$
       $= 60$
  a) Vans $= 60$
  b) Cars $= 6 \times 60$
      $= 360$

4 Multiplier $= 7000 \div 15 = 466.666$
  Adrian receives $2 \times 7000 \div 15 = £933.33$
  Penelope receives $5 \times 7000 \div 15 = £2333.33$
  Gladys receives $8 \times 7000 \div 15 = £3733.33$
  Check total: $933.33 + 2333.33 + 3733.33$
        $= 6999.99$
  (Notice 1p out due to rounding errors.)

Number: Working without a calculator (page 11)

1 Two significant figures
2 a) 5.76 **b)** 27.3 **c)** 3600
3 Estimate $= 2 \times 50$
       $= £100$
4 a) 0.043 **b)** $\frac{1}{7}$ **c)** 10
5 46

Number: Calculator methods (page 13)

1 a) 454.35 **b)** 36.96 **c)** 1.68
2 a) 0.02 or $\frac{1}{50}$ **b)** 1.33 **c)** $\frac{1}{9}$ or 0.11
3 a) 120 g **b)** $\frac{19}{20}$ **c)** £15.20
4 a) 847.51 **b)** 2.41 **c)** 1.03

Number: Solving problems (page 15)

1

Pack	Size	Price	Price per gram
Standard	500 g	£1.15	0.23p
Family	750 g	£1.59	0.212p
Special	1.2 kg	£2.49	0.2075p

Special is the best value for money.

2 Amount in dollars $= 500 \times 1.93$
$\qquad\qquad\qquad = \$965$
Amount left $= \$965 - \$784$
$\qquad\qquad\quad = \$181$
Received back $181 \div 1.93 = £93.78$

3 a) Estimate $= 50 \times 20$
$\qquad\qquad = £1000$
$£770.25$ is obviously much too little.

b) Increase $= £21.87 - 20.25$
$\qquad\qquad = £1.62$
$\qquad$ percentage increase $= \dfrac{1.62}{20.25} \times 100$
$\qquad\qquad\qquad\qquad\quad = 8\%$

4 Charge for 14 days $= 27.50 + 9 \times 4.50$
$\qquad\qquad\qquad\qquad = £68$

Algebra: Use of symbols (page 17)

1 $5s^3 - 10s$

2 $5a + 10 - 3a + 3 = 2a + 13$

3 $2x(2x - 1)$

4 $(4 \div 2)(a^6 \div a^2) = 2a^4$

Algebra: Linear equations (page 19)

1 $3m = 9 \times 4$
$\quad m = 36 \div 3$
$\quad m = 12$

2 $2y + 6 = 5y$
$\qquad 6 = 5y - 2y$
$\qquad 3y = 6$
$\qquad y = 6 \div 3$
$\qquad y = 2$

3 $4x + 8 + 6x - 4 = 14$
$\qquad 4x + 6x = 14 - 8 + 4$
$\qquad\quad 10x = 10$
$\qquad\qquad x = 1$

Algebra: Formulae (page 20)

1 $s = 9 \times 48 + \frac{1}{2} \times \left(-\frac{1}{4}\right) \times 48^2$
$\quad s = 432 - 288$
$\quad s = 144$

2 $2(C + 15) = F$
$\quad 2C + 30 = F$
$\qquad 2C = F - 30$
$\qquad\quad C = \dfrac{F - 30}{2}$ or $C = \dfrac{F}{2} - 15$

3 $5d + 3 = e$
$\quad 5d = e - 3$
$\quad\; d = \dfrac{e - 3}{5}$

Algebra: Inequalities (page 21)

1 $-2, -1, 0, 1, 2, 3, 4$

2 a) $8x > 25 - 5$
$\qquad 8x > 20$
$\qquad\; x > 20 \div 8$
$\qquad\; x > 2.5$

$-6 \;\; -5 \;\; -4 \;\; -3 \;\; -2 \;\; -1 \;\; 0 \;\; 1 \;\; 2 \;\; 3 \;\; 4 \;\; 5 \;\; 6$

b) $17 - 6 > 4x - 2x$
$\qquad 11 > 2x$
$\quad 11 \div 2 > x$
$\qquad\quad x < 5.5$

$-6 \;\; -5 \;\; -4 \;\; -3 \;\; -2 \;\; -1 \;\; 0 \;\; 1 \;\; 2 \;\; 3 \;\; 4 \;\; 5 \;\; 6$

Algebra: Trial and improvement (page 23)

1 a) Volume $= 4x \times x \times (x + 1) = 200\,\text{cm}^3$
$\qquad\qquad\qquad 4x^3 + 4x^2 = 200$
$\qquad\qquad\qquad\quad x^3 + x^2 = 50$

b)

x	$x^3 + x^2$		
3	36	too small	
4	80	too big	
3.5	55.125	too big	solution between 3 and 3.5
3.4	50.864	too big	
3.3	46.827	too small	solution between 3.3 and 3.4
3.35	48.818	too small	solution between 3.35 and 3.4
3.36	49.223	too small	
3.37	49.63	too small	
3.38	50.039	too big	solution between 3.37 and 3.38
3.375	49.834	too small	solution between 3.375 and 3.38

so answer is between 3.375 and 3.38 and to 3 s.f. the answer is $x = 3.38$

2 a) To show it has a root between 2 and 3, try each and show they have different signs.
Try $x = 2$ $\quad x^3 - 8x + 5 = -3$
Try $x = 3$ $\quad x^3 - 8x + 5 = 8$

b)

x	$x^3 - 8x + 5$		
2	-3	too small	
3	8	too big	solution between 2 and 3
2.5	0.625	too big	solution between 2 and 2.5
2.4	-0.376	too small	solution between 2.4 and 2.5
2.45	0.106	too big	solution between 2.4 and 2.45

so answer is between 2.45 and 2.4 and to 1 d.p. the answer is $x = 2.4$

Algebra: Sequences (page 24)

1 a) Difference $= 7$ so rule is $7n + k$. Term previous to 2 would be -5, so rule is $7n - 5$.

b) If $7n - 5 = 300$ then $7n = 305$ and $n = 305 \div 7$. Therefore $n = 43.57$ i.e. n is not a whole number therefore 300 is not in the sequence.

2 a)

t	1	2	3
c	4	8	12

$c = 4t$

b)

t	1	2	3
c	4	6	8

$c = 2t + 2$

Difference $= 2$ so rule is $c = 2t + k$. Previous term would be 2, so rule is $c = 2t + 2$.

Algebra: Plotting graphs (page 27)

1 a)

x	−1	0	1	2	3
y	4	1	0	1	4

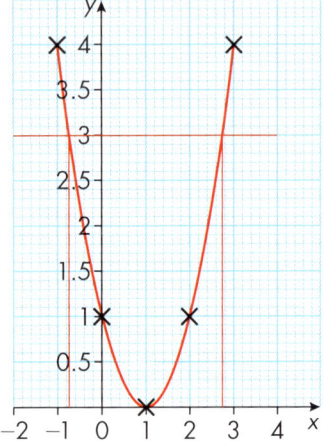

b) −0.7 and 2.7

2 a)

x	−1	0	1	2	3	4	5
y	−5	0	3	4	3	0	−5

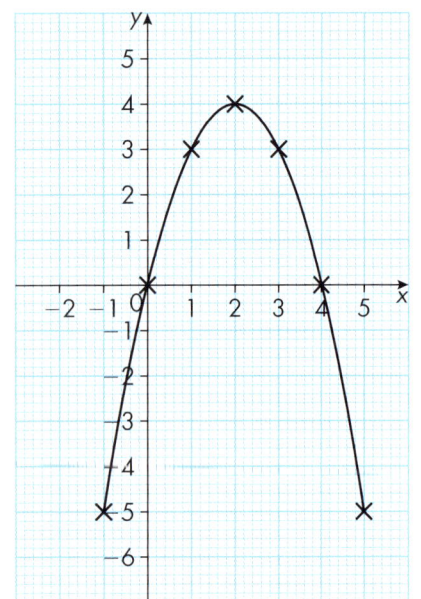

b) i) 2   **ii)** 0.6 and 3.4

Algebra: Interpreting graphs (page 28)

1 a) 4.5 km   **b)** He had stopped

c)

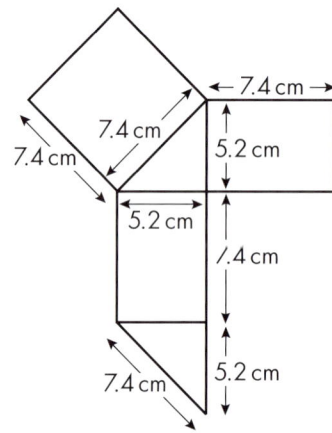

d) $9 \div 2 = 4.5$ km/h

Shape and space: Angles and two-dimensional shapes (page 33)

1 $x = 53°$ Corresponding angles are equal.
$y = 84°$ Angles in a triangle add to 180°.

2 a) and **b)**

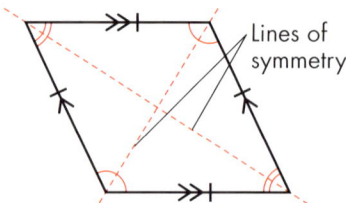

Lines of symmetry

c) Order of rotational symmetry is 2.

3 102°   Angles in a pentagon add to 540°.

4 a) Area of triangle $= \frac{1}{2} \times$ base $\times$ height
$= \frac{1}{2} \times 4.6 \times 5.0$
$= 11.5$ cm²

b) $a^2 = b^2 + c^2$
$= 4.6^2 + 5.0^2$
$= 46.16$
$a = \sqrt{46.16}$
$a = 6.8$ cm (to 1 d.p.)

Shape and space: Three-dimensional shapes (page 36)

1 Here is a possible net. Check the measurements on your net.

7.4 cm
7.4 cm
7.4 cm
7.4 cm
5.2 cm
5.2 cm
7.4 cm
5.2 cm
7.4 cm

2 a) 8
b) 12
c)

d) 15 cm³

3

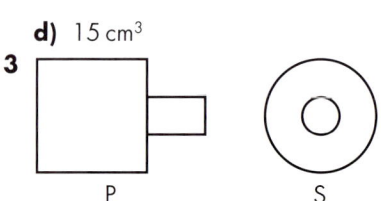

P   S

4 a) 42 cm³   **b)** 96 cm²

Shape and space: Transformations and coordinates (page 40)

1 a) Rotation through 90° clockwise about (0, 0)
   b) and **c)**

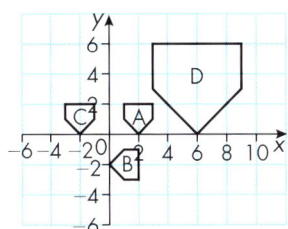

2 a) (0, 4, 0)   **b)** (0, 4, 2)
   c) (6, 4, 2)   **d)** (6, 2, 0)

Shape and space: Measures (page 42)

1 a)

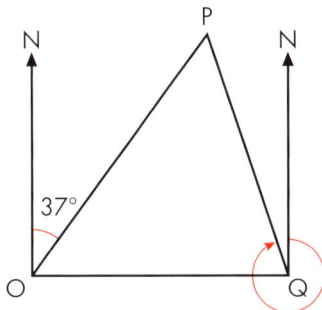

b) PQ measures 3.4 cm on the diagram, so is
   $3.4 \times 2 = 6.8$ km
c) 341°

2 a) i) 22 656 cm²
   ii) 2.2656 m² (or 2.27 m² to 2 d.p.)
   b) 353.5 cm to 354.5 cm and 63.5 cm to 64.5 cm

3 a) 639.6 cm³   **b)** 959.4 g

Shape and space: Constructions (page 46)

1 AX = 4.8 cm

2

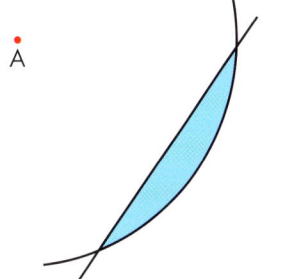

Diagram shown half-size.

Handling data: Identification and selection of data (page 48)

1 a) Wasim: Too few groups
   Abbie: Overlap, e.g. 10 in two groups

b)

Cars		Frequency
0–9	|||	3
10–19	┼┼┼┼ ┼┼┼┼	10
20–29	┼┼┼┼ ┼┼┼┼ ||	12
30–39	┼┼┼┼		6
40–49	|||	3
50–59	||	2

Handling data: Processing and representing data (page 51)

1 a) Ordered

```
1 |
2 | 1  4                Key: 2|1 means 21
3 | 4
4 | 3  4  8
5 | 5  6  8
6 | 1  4  5  9
7 | 2  2  3  [4]  6  6  8  8  9
8 | 1  1  1  2  2  9
9 | 1  2  5  8  9
```

b) 74

2 Mean
$$= \frac{45 \times 14 + 46 \times 25 + 47 \times 32 + 48 \times 19 + 49 \times 8 + 50 \times 2}{14 + 25 + 32 + 19 + 8 + 2}$$
$$= 46.88$$

Handling data: Comparing distributions (page 52)

1 Average leaf length about the same
(Tree A Median = 7.2 cm, Tree B Median = 7.3 cm).
Length of leaves more spread out for Tree A
(range: Tree A 4.2 cm, Tree B 2.3 cm).

Handling data: Probability (page 54)

1 $72 + 15 + 28 + 33 + 12 = 160$
   a) i) $\frac{72}{160} = \frac{9}{20}$
   ii) $\frac{12}{160} = \frac{3}{40}$
   iii) $160 - 15 = 145$ so probability is $\frac{145}{160} = \frac{29}{32}$

   b) No. People travel for different reasons at different
   times of the day so the proportions of vehicles will
   be different.

2 a) P(Blue) $= 1 - (0.2 + 0.45) = 0.35$
   b) $0.45 \times 200 = 90$

Section 2

Number: Powers and roots (page 57)

1 a) $\frac{1}{9^2} = \frac{1}{81}$ **b)** 1   **c)** $\sqrt[3]{27} = 3$

2 a) $(3 \times 6) \times (10^4 \times 10^3) = 18 \times 10^7 = 1.8 \times 10^8$
   b) $1.2 \times 10^8 \div 10^{12} = 1.2 \times 10^{-4}$ terawatts

Number: Fractions and decimals (page 59)

1 a) $\dfrac{8}{\cancel{5}_1} \times \dfrac{\cancel{20}^4}{9} = \dfrac{32}{9} = 3\dfrac{5}{9}$

b) $(1 + 2) + \left(\dfrac{1}{4} + \dfrac{3}{5}\right) = 3 + \dfrac{1 \times 5 + 3 \times 4}{20} = 3\dfrac{17}{20}$

2 $\dfrac{2xy}{x + y} = \dfrac{2 \times \frac{2}{5} \times \frac{2}{7}}{\frac{2}{5} + \frac{2}{7}} = \dfrac{\frac{8}{35}}{\frac{24}{35}} = \dfrac{\cancel{8}^1}{_1\cancel{35}} \times \dfrac{\cancel{35}^1}{\cancel{24}_3} = \dfrac{1}{3}$

Number: Percentages (page 60)

1 a) $A \times 0.8 = 10\,240$ so $10\,240 \div 0.8 = £12\,800$

b) $10\,240 \div (0.80)^3 = £20\,000$

Number: Written methods (page 61)

1 a) i) $\sqrt{36} \times \sqrt{2} = 6\sqrt{2}$

ii) $\sqrt{5} \times \sqrt{4} \times \sqrt{5} \times \sqrt{3} = 5 \times 2 \times \sqrt{3} = 10\sqrt{3}$

iii) $\dfrac{\sqrt{25}\,\cancel{\sqrt{2}} \times \cancel{\sqrt{9}}\,\sqrt{3}}{\cancel{\sqrt{9}}\,\cancel{\sqrt{2}}} = 5\sqrt{3}$

b) $(5 + \sqrt{7})(5 + \sqrt{7}) = 25 + 5\sqrt{7} + 5\sqrt{7} + \sqrt{7}\sqrt{7}$
$\qquad = 32 + 10\sqrt{7}$

$a = 32, \ b = 10$

2 $14 \times \dfrac{250}{80} = 43.75$ litres

3 $£9500 \times \dfrac{3}{8} = £3562.50$

Number: Calculator methods (page 63)

1 a) 1.897×10^{-5} **b)** 6×10^{-5}

2 a) 6.43×10^{10} **b)** 1.13×10^{32}

Algebra: Expansions, factors and indices (page 65)

1 $12p^2q - 15pq^2 = 3 \times 4 \times p \times p \times q - 3 \times 5 \times p \times q$
$\qquad \times q = 3pq(4p - 5q)$

2 a) $3 \times 4 \times a^{2+1} \times b^{1+3} = 12a^3b^4$

b) $a^{3-2}b^{5-3} = ab^2$

3 $\dfrac{m^{3+4} \times n^{3+2}}{m^5n} = \dfrac{m^7n^5}{m^5n} = m^{7-5}n^{5-1} = m^2n^4$

4 a) $(3 - x)(3 - x) = 9 - 3x - 3x + x^2 = 9 - 6x + x^2$

b) $3x^2 + 12x - 2x - 8 = 3x^2 + 10x - 8$

Algebra: Formulae (page 66)

1 $\dfrac{P}{5} = \sqrt{V}$

$\dfrac{P^2}{25} = V$

$V = \dfrac{P^2}{25}$

2 $\dfrac{1}{3}\pi r^2 h = V$

$\pi r^2 h = 3V$

$r^2 = \dfrac{3V}{\pi h}$

$r = \sqrt{\dfrac{3V}{\pi h}}$

3 a) $de - 5d = 3$

$d(e - 5) = 3$

$d = \dfrac{3}{e - 5}$

b) $3d - 7 = e(4 + 5d)$

$3d - 7 = 4e + 5de$

$3d - 5de = 4e + 7$

$d(3 - 5e) = 4e + 7$

$d = \dfrac{4e + 7}{3 - 5e}$

Algebra: Direct and inverse proportion (page 68)

1 $R \propto s^2$

$R = ks^2$

$100 = k300^2$

$k = \dfrac{100}{90\,000} = \dfrac{1}{900}$

So $R = \dfrac{1}{900}s^2$

a) $R = \dfrac{1}{900} \times 600^2 = 400\,\text{N}$

b) $200 = \dfrac{1}{900}s^2$

$s^2 = 180\,000$

$s = \sqrt{180\,000}$

$s = 424\,\text{m/s}$

2 a) $10 \times 4 = 40$ **b)** $10 \times 4^2 = 160$

c) $10 \div 4 = 2.5$

3 a) $y \propto \dfrac{1}{x^2}$

$y = \dfrac{k}{x^2}$

$9 = \dfrac{k}{2^2}$

$k = 36$

So $y = \dfrac{36}{x^2}$

b) $1 = \dfrac{36}{x^2}$

$x^2 = 36$

$x = \pm 6$

Algebra: Gradient and equations of straight lines (page 70)

1 a) gradient, $m = -2$ y-intercept, $c = 4$

b) $m = -2, c = -1$

So $y = -2x - 1$

2 $y = -x + 1$ so $m = -1$

Parallel to $x + y = 1$:

gradient, $m = -1$

$y = -x + c$

since line goes through $(1, 1)$

$1 = -1 + c$

$c = 2$

so line is $y = -x + 2$

Perpendicular to $x + y = 1$:

gradient, $m = \dfrac{-1}{-1} = 1$

$y = x + c$

since line goes through $(1, 1)$

$1 = 1 + c$

$c = 0$

so line is $y = x$

Algebra: Quadratic functions and equations (page 73)

1 a) $5(x^2 - 4) = 5(x + 2)(x - 2)$
  b) i) $(x - 8)(x - 1)$ **ii)** $x = 8$ or 1

2 $a = 2$ $b = -38$ $c = 45$
$$x = \frac{-(-38) \pm \sqrt{(-38)^2 - 4 \times 2 \times 45}}{2 \times 2}$$
$$x = \frac{38 \pm \sqrt{1084}}{4}$$
$x = 17.73$ or 1.27 to 2 d.p.

3 a) $(x - 6)^2 - 36 + 2 = (x - 6)^2 - 34$
  b) -34
  c) $(x - 6)^2 - 34 = 0$
$$(x - 6)^2 = 34$$
$$x - 6 = \pm\sqrt{34}$$
$$x = 6 \pm \sqrt{34}$$
$$x = 11.83 \text{ or } 0.17 \text{ to 2 d.p.}$$

Algebra: Simultaneous equations (page 75)

1 a)
$$5x + 4y = 13 \qquad (1)$$
$$3x + 8y = 5 \qquad (2)$$
For example, using Elimination Method.
$(1) \times 2$ $10x + 8y = 26$ (3)
$(3) - (2)$ $7x = 21$
$$x = 3$$
Substitute $x = 3$ in (1)
$$15 + 4y = 13$$
$$4y = -2$$
$$y = -0.5$$
so $x = 3$; $y = -0.5$

b)
$$4x + 3y = 5 \qquad (1)$$
$$2x + y = 1 \qquad (2)$$
For example, using Substitution Method.
Rearrange (2) $y = 1 - 2x$ (3)
Substitute for y in (1)
$$4x + 3(1 - 2x) = 5$$
$$4x + 3 - 6x = 5$$
$$-2x = 2$$
$$x = -1$$
Substitute for x in (3)
$$y = 1 - 2 \times (-1)$$
$$y = 3$$
So $x = -1$; $y = 3$

c)
$$2x - 3y = 9 \qquad (1)$$
$$5x + 2y = -25 \qquad (2)$$
For example, using Elimination Method.
$(1) \times 2$ and $(2) \times 3$
$$4x - 6y = 18 \qquad (3)$$
$$15x + 6y = -75 \qquad (4)$$
$(3) + (4)$
$$19x = -57$$
$$x = -3$$
Substitute for x in (1)
$$2 \times (-3) - 3y = 9$$
$$-6 - 3y = 9$$
$$-3y = 15$$
$$y = -5$$
So $x = -3$, $y = -5$

2 Substitute the linear equation into the quadratic equation.
$$x^2 + (3x - 1)^2 = 12$$
$$x^2 + (3x - 1)(3x - 1) = 12$$
$$x^2 + 9x^2 - 3x - 3x + 1 = 12$$
$$10x^2 - 6x - 11 = 0$$
$a = 10$, $b = -6$, $c = -11$
$$x = \frac{6 \pm \sqrt{(-6)^2 - 4(10)(-11)}}{2 \times 10}$$
$$= \frac{6 \pm \sqrt{476}}{20} = \frac{6 + 21.817}{20} \text{ or } 6\frac{-21.817}{20}$$
$$= 1.39 \text{ or } -0.791$$
$x = 1.39$, $y = 3 \times 1.39 - 1 = 3.17$
$x = -0.791$ $y = 3 \times -0.791 - 1 = -3.373$
Coordinates $(1.4, 3.2)$ and $(-0.8, -3.4)$
correct to 1 d.p.

Algebra: Algebraic fractions (page 77)

1 a) $\dfrac{\cancel{(x - 3)}(x + 3)}{\cancel{(x - 3)}(x + 2)} = \dfrac{x + 3}{x + 2}$

  b) Multiply through by $(x + 1)(3x + 1)$
$$12(x + 1) - 5(3x + 1) = (x + 1)(3x + 1)$$
$$12x + 12 - 15x - 5 = 3x^2 + 4x + 1$$
$$0 = 3x^2 + 7x - 6$$
$$(3x - 2)(x + 3) = 0$$
$$x = \frac{2}{3} \text{ or } -3$$

2 a) $\dfrac{2x - 1(x - 3)}{x(x - 3)} = \dfrac{2x - x + 3}{x(x - 3)} = \dfrac{x + 3}{x(x - 3)}$

  b) $\dfrac{2(x - 2) + 3(x + 1)}{(x + 1)(x - 2)} = \dfrac{2x - 4 + 3x + 3}{(x + 1)(x - 2)}$
$$= \dfrac{5x - 1}{(x + 1)(x - 2)}$$

3 a) $\frac{48}{12} = 4$ hours
  b) Return time $= \frac{48}{8} = 6$ hrs
  so average speed $= \dfrac{48 + 48}{6 + 4} = \dfrac{96}{10} = 9.6$ km/h
  c) i) $\dfrac{48}{x} + \dfrac{48}{x - 5} = 8$
  Divide through by 8: $\dfrac{6}{x} + \dfrac{6}{x - 5} = 1$
  ii) Multiply through by $x(x - 5)$
$$6(x - 5) + 6x = x(x - 5)$$
$$6x - 30 + 6x = x^2 - 5x$$
$$0 = x^2 - 17x + 30$$
$$(x - 2)(x - 15) = 0$$
$$x = 2 \text{ or } 15$$
  iii) But $x \neq 2$ otherwise return journey speed would be negative so $x = 15$ km/h; so average speed for return journey $= 15 - 5 = 10$ km/h

Algebra: Graphical solution of simultaneous equations and linear inequalities (page 79)

1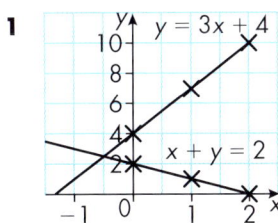

$x = -0.5$, $y = 2.5$

2 $y = 0$ is the x-axis.

$x + y = 5$	passes through $(0, 5)$ and $(5, 0)$.
$y = 2x - 1$	passes through $(0, -1)$ and $(1, 1)$ and $(2, 3)$.
$y \leqslant 0$	For $(0, 1)$, $y \leqslant 0$ is false.

Shade the $(0, 1)$ (unwanted) side.

$x + y \leqslant 5$ For $(0, 0)$, $x + y \leqslant 5$ is true.

Shade the opposite (unwanted) side to the origin.

$y \leqslant 2x - 1$ For $(0, 0)$, $y \leqslant 2x - 1$ is false.

Shade the origin (unwanted) side.

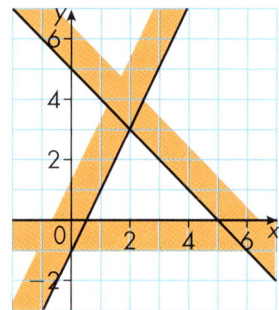

3 Choose $(0, 1)$ in the region.

For	$x + y = 3$	$0 + 1 \leqslant 3$	so	$x + y \leqslant 3$
For	$y = x$	$1 \geqslant 0$	so	$y \geqslant x$
For	$x = -2$	$0 \geqslant -2$	so	$x \geqslant -2$

Algebra: Graphs of functions (page 83)

1 a)

x	−5	−4	−3	−2	−1
y	−0.4	−0.5	−0.67	−1	−2

	1	2	3	4	5
	2	1	0.67	0.5	0.4

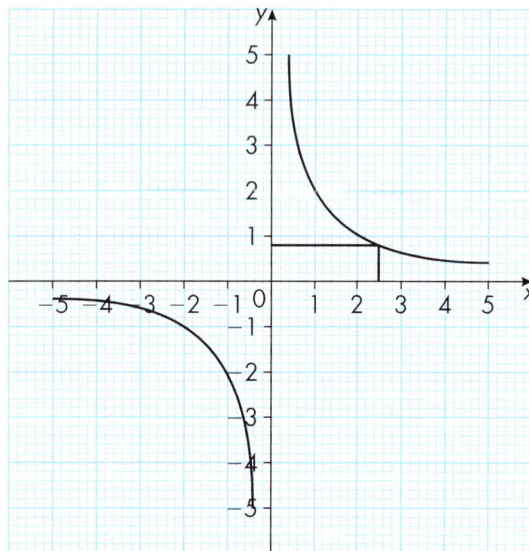

b) $x = 2.5$

2 a)

x	−3	−2	−1	0	1	2	3
y	−16	−1	2	−1	−4	−1	14

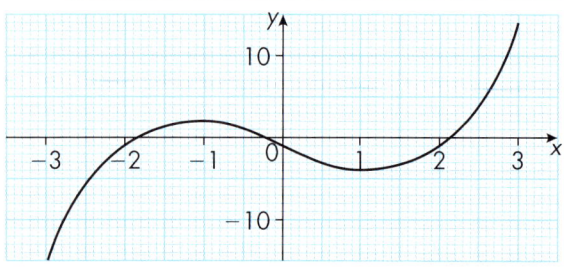

b) −2 to −1.7, −0.4 to −0.2, 2.0 to 2.2

c) $x^3 - 6x - 3 = 0$

$x^3 - 4x - 1 - 2x - 2 = 0$

$x^3 - 4x - 1 = 2x + 2$

Draw $y = 2x + 2$

$y = 2x + 2$ drawn, −2.4 to −2.1, −0.6 to −0.3, 2.6 to 2.8

Algebra: Transformations and functions (page 86)

1 a)

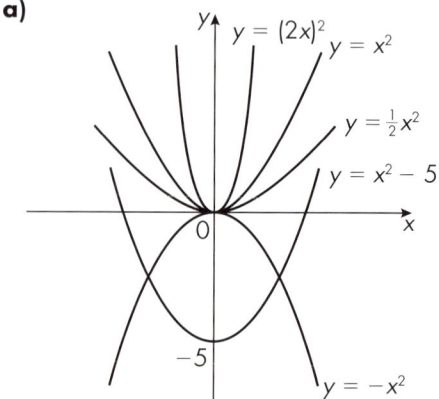

b) i) Reflection in x-axis

ii) Translation $\begin{pmatrix} 0 \\ -5 \end{pmatrix}$

iii) One-way stretch parallel to the y-axis with scale factor $\frac{1}{2}$

iv) One-way stretch parallel to the x-axis with scale factor $\frac{1}{2}$

2

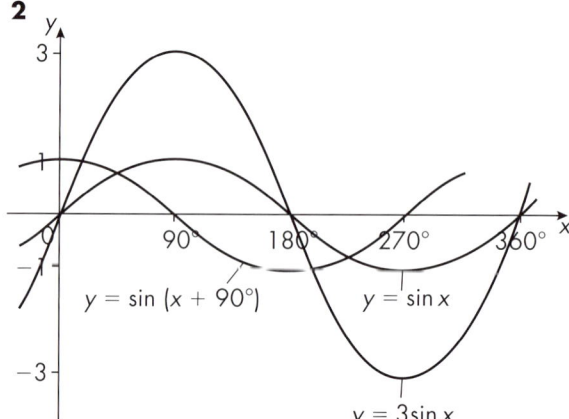

Algebra: Graphs in real life (page 87)

1 a)

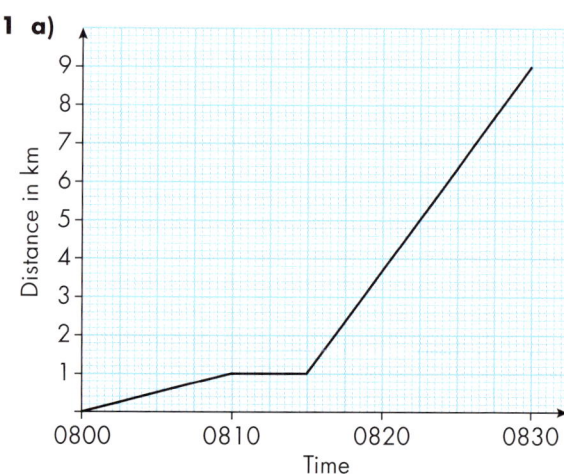

b) $3\frac{2}{3}$ km or 3.7 km.

2 e.g. A car accelerates and then travels at a steady speed and then stops suddenly (crashes etc).

3 a) 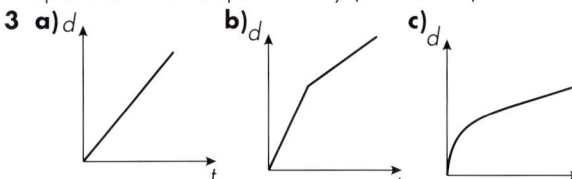 **b)** **c)**

Shape and space: Properties of triangles and other shapes (page 90)

1 In triangle PCX and triangle ARX
   PX = RX (sides of a square)
   CX = AX (sides of a square)
   Angle CXP = angle RXA (both 90° + angle CXR).
   So triangle PCX = triangle RAX (SAS)
   Therefore angle PCX = angle RAX (corresponding angles).

Shape and space: Pythagoras and trigonometry (page 92)

1 a) i) $\sqrt{5^2 + 8^2} = 9.43$ cm
   ii) $\sqrt{7^2 + 8^2} = 10.63$ cm
   iii) $\sqrt{5^2 + 7^2} = 8.60$ cm
   b) $9.43^2 = 10.63^2 + 8.60^2 - 2 \times 10.63 \times 8.60$
   $\times \cos ACB$
   $\cos ACB = \dfrac{10.63^2 + 8.60^2 - 9.43^2}{2 \times 10.63 \times 8.60}$
   $ACB = 57.6°$ 1 d.p.

Shape and space: Properties of circles (page 94)

1 $x = 36°$ (angles in the same segment)
   $y = 72°$ (angle at centre = twice angle at circumference)
   $z = 62°$ (radius at right angles to tangent)

Shape and space: Three-dimensional shapes (page 95)

1 $\frac{1}{3} \times \pi \times 3.5^2 \times 11 + \frac{1}{2}(\frac{4}{3} \times \pi \times 3.5^3)$
   $= 230.9$ cm^3
2 a) $40 \div 1.25 = 32$ cm
   b) $32\,000 \times 1.25^3 = 62\,500$ cm^3

Shape and space: Transformations, coordinates and vectors (page 98)

1 Enlargement with the centre (0, 0) and scale factor −1.
2 a) B is (−2 + 6, 1 + 5) or (4, 6)
   M is $\left(\dfrac{-2 + 4}{2}, \dfrac{1 + 6}{2}\right)$ or (1, 3.5)
   N is $\left(\dfrac{-2 + 2}{2}, \dfrac{1 + -2}{2}\right)$ or (0, −0.5)

b) $\overrightarrow{BC} = \begin{pmatrix} 2 - 4 \\ -2 - 6 \end{pmatrix} = \begin{pmatrix} -2 \\ -8 \end{pmatrix}$ or from diagram
   $\overrightarrow{MN} = \begin{pmatrix} 0 - 1 \\ -0.5 - 3.5 \end{pmatrix} = \begin{pmatrix} -1 \\ -4 \end{pmatrix}$ or from diagram
c) MN is parallel to BC and BC = 2 × MN

Shape and space: Measures (page 99)

1 For maximum a need maximum $(v - u)$ and minimum t
   Max. $(v - u) = 30.35 - 17.35 = 13$
   Min. t $= 2.55$
   So maximum $a = \dfrac{13}{2.55} = 5.1$ (1 d.p.)

Shape and space: Arcs and sectors of circles (page 100)

1 $\dfrac{150}{360} \times 2\pi r = 20$
   $r = \dfrac{20 \times 360}{150 \times 2 \times \pi}$
   $r = 7.64$ cm (2 d.p.)

Handling data: Processing and representing data (page 104)

1 a) i) [9], 22, 43, 60, 70, 78, 80
   ii) and e)

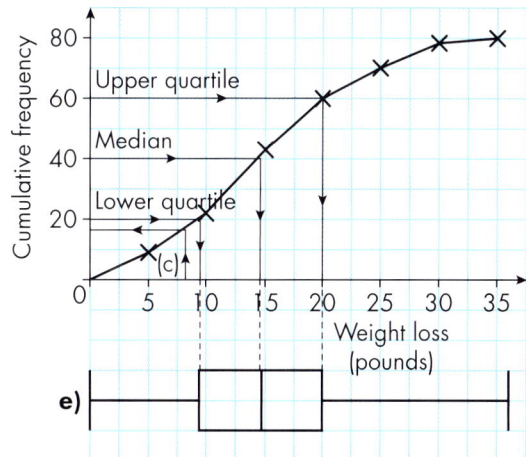

b) i) 14 to 14.8 pounds
   ii) 10.5 to 11 is acceptable
c) 63 to 65 is acceptable
d) $84 \div 0.7 = 120$ kg

2 Frequency densities:
   $\frac{14}{5} = 2.8$; $\frac{41}{10} = 4.1$; $\frac{59}{10} = 5.9$; $\frac{70}{20} = 3.5$; $\frac{16}{30} = 0.53$

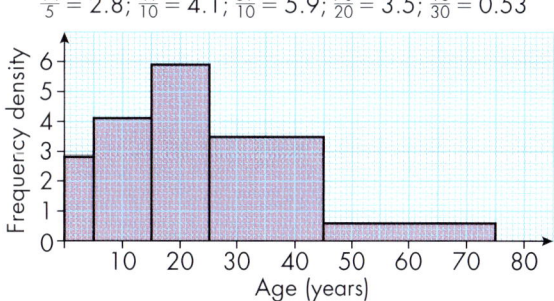

3 a)

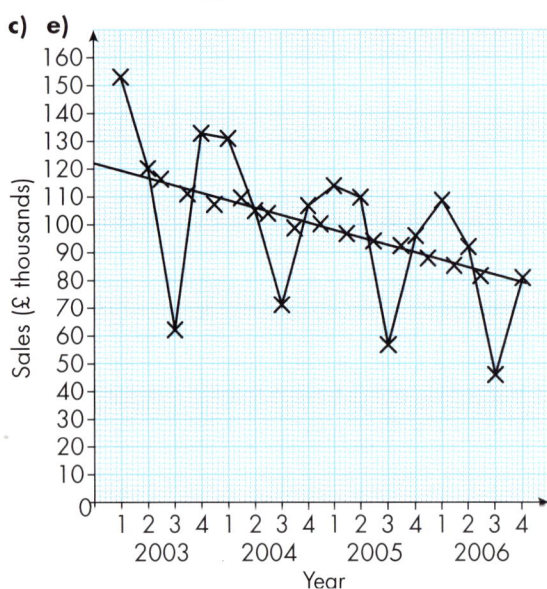

b) $a = \dfrac{153 + 120 + 62 + 133}{4} = 117$

$b = \dfrac{120 + 62 + 133 + 131}{4} = 111.5$

c) e)

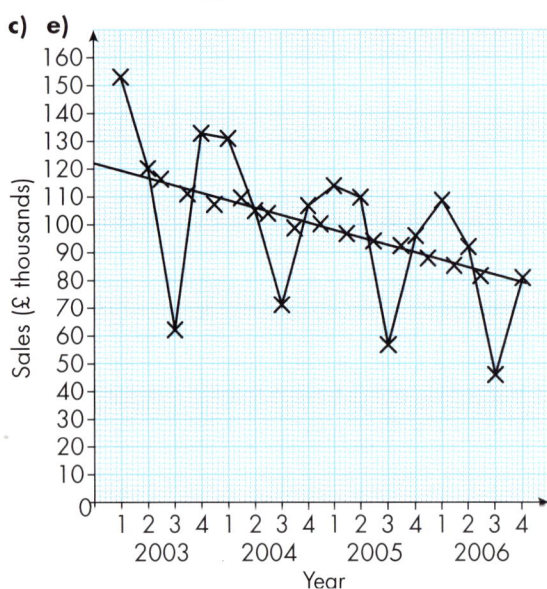

d) General trend: steadily lower
Quarterly variation: e.g. highest always 1st quarter, lowest always 3rd quarter

f) Approx 80.25

g) $80.25 = \dfrac{92 + 46 + 81 + x}{4}$

$x = 4 \times 80.25 - (92 + 46 + 81)$

$x = 102$ So sales is £102 000

4 a)

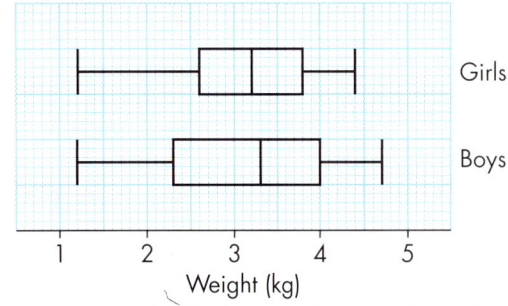

b) e.g. boys tend to be heavier, the range for girls is less, the interquartile range for the girls is less, 25% of the boys weigh less than 2.3 kg and 25% of the girls weigh less than 2.6 kg so there are

more boys with low weights, girls' weights are more consistent i.e. 50% of the girls have a weight between 2.6 kg and 3.8 kg but for the boys 50% weigh between 2.3 kg and 4.0 kg.

Handling data: Probability (page 107)

1 a)

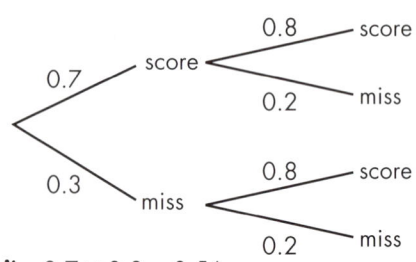

b) i) $0.7 \times 0.8 = 0.56$

ii) $0.7 \times 0.2 + 0.3 \times 0.8 = 0.14 + 0.24$
$= 0.38$

2 a)

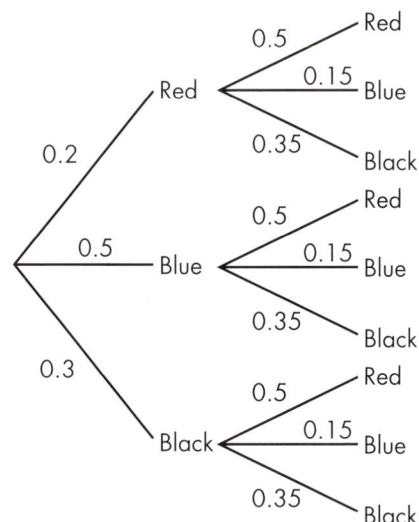

b) $0.2 \times 0.5 = 0.1$ or 10%

c) $0.2 \times 0.5 + 0.5 \times 0.15 + 0.3 \times 0.35 = 0.28$ or 28%

3 a) $P = \dfrac{2}{12} = \dfrac{1}{6}$

b) $\dfrac{5}{30} \times \dfrac{4}{29} = \dfrac{1}{6} \times \dfrac{4}{29} = \dfrac{1}{3} \times \dfrac{2}{29} = \dfrac{2}{87}$

c) RRL + RLR + LRR
$\dfrac{10}{12} \times \dfrac{9}{11} \times \dfrac{2}{10} + \dfrac{10}{12} \times \dfrac{2}{11} \times \dfrac{9}{10} + \dfrac{2}{12} \times \dfrac{10}{11} \times \dfrac{9}{10} = \dfrac{3}{22} \times 3$
$= \dfrac{9}{22}$

Handling data: Sampling (page 109)

1 a) Advantage: cheaper, quicker
Disadvantage: may not be representative

b) May be biased, e.g. may be in friendship or year groups or could have all just got off school buses

c) Number all students. Select random numbers, e.g. raffle tickets, random number generator, until she has 10%

d) Ensures all groups represented, e.g. each stratum is a year group

2 a) Strata different areas, council tax bands, flats, 2 bedroom, 3 bedroom etc. Then select 10% sample from each stratum

b) Ensures all types of household represented

c) May not give a representative sample, only gets those in phone book not those with no phone, ex-directory etc

Integers

Number

Integers

Prime factors ☐
Highest common factor (HCF) ☐
Lowest common multiple (LCM) ☐
Indices ☐

Fractions

Mixed numbers and improper fractions ☐
Adding and subtracting functions ☐
Multiplying fractions ☐
Dividing fractions ☐

Percentages

Finding A as a percentage of B ☐
Percentage increases or decreases ☐
Repeated percentage change ☐

Ratio

Writing as a ratio ☐
Mixing in a ratio ☐
Sharing in a given ratio ☐
Direct proportion ☐

Working without a calculator

Multiplying and dividing decimals ☐
Rounding to a given number of significant
   figures ☐
Estimating ☐
Reciprocals ☐
Using place value ☐

Calculator methods

Getting to know your calculator ☐
Reciprocals ☐
Powers and roots ☐

Solving problems

Some common money problems ☐
Compound measures ☐
Rounding your answers and checking your
   work ☐

Algebra

Use of symbols

Multiplying brackets by single terms ☐
Common factors ☐
Multiplying out two sets of brackets ☐
Indices ☐

Linear equations

Brackets in equations ☐
Unknown quantity on both sides of the
   equation ☐
More than one set of brackets ☐
Fractions in equations ☐

Formulae

Substituting in formulae ☐
Writing your own formulae ☐
Rearranging formulae ☐

Inequalities

Trial and improvement

Sequences

Common sequences ☐
General rules for sequences ☐

Plotting graphs

Straight lines ☐
Quadratic graphs ☐
Using graphs to solve quadratic equations ☐

Interpreting graphs

Graphs in real situations ☐

Shape and space

Angles and two-dimensional shapes

Basic angle facts ☐
Angles with parallel lines ☐
Quadrilaterals ☐
Polygons ☐
Pythagoras' theorem ☐
Areas and perimeters ☐

Three-dimensional shapes

The net of a solid ☐
Isometric drawing ☐
Plans and elevations ☐
Volume and surface area of prisms ☐

Transformations and coordinates

Reflection ☐
Rotation ☐
Enlargement ☐
Translations ☐
Coordinates ☐

Measures

Bearings and scale drawings ☐
Changing units ☐
Bounds of measurement ☐
Compound measures ☐

Constructions

Constructing a triangle ☐
Other standard ruler and compass
constructions ☐
Loci ☐

Handling data

Identification and selection of data

Primary and secondary data ☐
Grouped data and class intervals ☐
Surveys ☐

Processing and representing data

Frequency diagrams and polygons ☐
Scatter diagrams and lines of best fit ☐
Mean from a frequency distribution ☐
Stem-and-leaf diagrams and median ☐
Comparing distributions ☐

Probability

Basic probability ☐
Relative frequency ☐

Section 2

Number

Powers and roots

Indices which are integers ☐
Exponential growth and decay ☐
Multiplying and dividing with indices ☐
Indices which are fractions ☐
Standard index form ☐

Fractions and decimals

Multiplying and dividing fractions ☐
Increasing or decreasing by a fraction ☐
Finding an amount before an increase or
   decrease ☐
Recurring and terminating decimals ☐
Changing a recurring decimal to a fraction ☐

Percent...

Finding
   inc... or decrease ☐

V...en methods

...ct proportion ☐
...erse proportions ☐
...urds ☐

Calculator methods

Standard form ☐
Trigonometric keys ☐

Algebra

Expansions, factors and indices

Multiplying two brackets ☐
Common factors ☐
Indices ☐

Formulae

Rearranging formulae ☐
Powers of the subject ☐
Subject twice in formula ☐

Direct and inverse proportion

Direct proportion ☐
Inverse proportion ☐

Gradient and equations of straight lines

Gradient and y-intercept ☐
The general equation of a straight
   line $y = mx + c$ ☐
Parallel and perpendicular lines ☐

Quadratic functions and equations

Factorising quadratics
Difference of two squares ☐
Solving quadratic equati...
☐
   ...o not
...tories ☐

Simultaneous equations

Elimination method ☐
Substitution method ☐
Simultaneous equations: one linear, one
   quadratic ☐
Simultaneous equations: one linear, one a
   circle ☐

Algebraic fractions

Simplifying fractions ☐
Solving equations with fractions ☐

Graphical solution of simultaneous equations and linear inequalities

Graphical solution of simultaneous equations ☐
Graphical solution of a set of linear
   inequalities ☐

Graphs of functions

Cubic graphs

Reciprocal and exponential graphs

Trigonometrical graphs

Transformations and functions

Function notation

Translations

Reflections

Stretches

Graphs in real life

Graphs in real situations

Drawing graphs

Shape and space

Properties of triangles and other shapes

Congruent triangles

Similar triangles

Pythagoras and trigonometry

Pythagoras' theorem and trigonometry

Sine and cosine rules

Properties of circles

Sectors and segments

Tangent properties of circles

Chord properties of circles

Angle properties of circles

Three-dimensional shapes

Pyramids, cones and spheres

Volume and surface area of similar shapes

Transformations, coordinates and vectors

Enlargement

Finding the single transformation equivalent

to two given transformations

Calculating with coordinates

Length, area and volume

Vector geometry

Measures

Upper and lower bounds of combined

measurements

Arcs and sectors of circles

Arc length and area of a sector of a circle

Handling data

Processing and representing data

Histograms

Time series and moving averages

Cumulative frequency graphs, median, quartiles,

interquartile range and box plots

Comparing distributions

Probability

Combining probabilities

Conditional probabilities

Sampling

Simple random sampling

Stratified sampling

Index